Twenty-Five Centuries of Technological Change

An Historical Survey

FUNDAMENTALS OF PURE AND APPLIED ECONOMICS

Fundamentals of Pure and Applied Economics is an international series of titles divided by discipline into sections. A listing of sections and their editors and a list of published titles may be found at the back of this volume.

Twenty-Five Centuries of Technological Change

An Historical Survey

Joel Mokyr
Departments of Economics and History, Northwestern University, Evanston, Illinois, USA

A volume in the Economics of Technological Change section

edited by

F. M. Scherer
Swarthmore College, Pennsylvania, USA

Routledge
Taylor & Francis Group
LONDON AND NEW YORK

First published 1990 by Harwood Academic Publishers

Published 2021 by Routledge
2 Park Square, Milton Park, Abingdon, Oxon OX14 4RN
605 Third Avenue, New York, NY 10017

Routledge is an imprint of the Taylor & Francis Group, an informa business

Library of Congress Cataloging-in-Publication Data

Mokyr, Joel.
Twenty-five centuries of technological change
An historical survey/Joel Mokyr.
p. cm.—(Fundamentals of pure and applied economics; v. 35)
"A volume in the economics of technological change section edited by F. M. Scherer."
Bibliography: p.
Includes index.
ISBN 3-7186-4936-5
1. Technological innovations—Economic aspects—History.
2. Economic history. I. Scherer, F. M. (Frederic M.) II. Title.
III. Title: 25 centuries of technological change. IV. Series.
HC79.T4M65 1989 89-7409
609—dc20 CIP

ISBN 13: 978-3-7186-4936-5 (pbk)

CONTENTS

Introduction to the Series

Drawing on a personal network, an economist can still relatively easily stay well-informed in the narrow field in which he works, but to keep up with the development of economics as a whole is a much more formidable challenge. Economists are confronted with difficulties associated with the rapid development of their discipline. There is a risk of "balkanization" in economics, which may not be favorable to its development.

Fundamentals of Pure and Applied Economics has been created to meet this problem. The discipline of economics has been subdivided into sections (listed at the back of this volume). These sections include short books, each surveying the state of the art in a given area.

Each book starts with the basic elements and goes as far as the most advanced results. Each should be useful to professors needing material for lectures, to graduate students looking for a global view of a particular subject, to professional economists wishing to keep up with the development of their science, and to researchers seeking convenient information on questions that incidentally appear in their work.

Each book is thus a presentation of the state of the art in a particular field rather than a step-by-step analysis of the development of the literature. Each is a high-level presentation but accessible to anyone with a solid background in economics, whether engaged in business, government, international organizations, teaching, or research in related fields.

Three aspects of *Fundamentals of Pure and Applied Economics* should be emphasized:

—First, the project covers the whole field of economics, not only theoretical or mathematical economics.

—Second, the project is open-ended and the number of books is not predetermined. If new and interesting areas appear, they will generate additional books.

—Last, all the books making up each section will later be grouped to constitute one or several volumes of an Encyclopedia of Economics.

The editors of the sections are outstanding economists who have selected as authors for the series some of the finest specialists in the world.

J. Lesourne *H. Sonnenschein*

Twenty-Five Centuries of Technological Change: An Historical Survey

JOEL MOKYR

Departments of Economics and History, Northwestern University

1. INTRODUCTION

One of the most pervasive and pernicious half-truths economists teach their students is the hackneyed aphorism that there is no such thing as a free lunch. It is the purpose of this essay to highlight the greatest counterexample to this statement.[1] Economic History is full of examples of free lunches, as well as (more frequently) very cheap lunches. At the same time, there are endless instances of very expensive meals that ended up inedible and in some cases lethal. Progress and growth are the subjects of this essay, but human history is laden with waste and inefficiency; the past has rarely been an example of economic efficiency. To phrase it differently, technological change is primarily the study of outward shifts of the production possibility frontier. Yet often societies have not been on the frontier, but at a point far from any economic optimality.

This monograph has benefitted throughout its long period of creation from the wisdom and wit of F. Michael Scherer. The additional comments of William Baumol, Reuven Brenner, Julia Burns, Charles Calomiris, Jan De Vries, Karl deSchweinitz, Stefano Fenoaltea, Jack Goldstone, C. Knick Harley, Dan Headrick, Jonathan R. T. Hughes, David Hull, Eric L. Jones, William McNeill, Cormac Ó Gráda, and William N. Parker, on earlier drafts have resulted in innumerable improvements. Mrs Barbara Karni provided her usual peerless editorial expertise. All remaining errors are mine and my sources' alone.

[1] For a similar statement, rare amongst economists, see Kamien and Schwartz (1982), p. 216. Kamien and Schwartz, too, regard technological change as a "trick" that makes it possible to avoid a choice and have "both" when faced with Samuelson's famous query "which one." Their analysis is cast largely in terms of a modern market in which research and development are embarked upon in a systematic way.

Roughly speaking, economic growth can occur as a result of four basic causes:

a) Growth that derives from capital deepening, or Solovian Growth (named after Robert Solow, who laid the foundation of the theory of modern economic growth). Because output per capita depends on the capital–labor ratio, net capital formation at a rate faster than population growth leads to economic growth, defined as an increase in output per capita. This type of growth involves no free lunch: in principle, all future benefits are paid for by abstinence in the present.

b) Growth that derives from commercial expansion leading to a more efficient allocation of resources. Any economist can show how the emergence of exchange (of goods or factors of production) can be beneficial to all partners involved, whether the gains are from international or local trade. A finer division of labor leads to productivity growth through specialization and the adaptation of skills to tasks. This process may be termed "Smithian Growth" (following Parker (1984)). Abstracting from the cost of transacting, the emergence of Smithian growth is a good example of a free lunch. It is, however, not the kind of free lunch I shall be much concerned with below.

c) Growth that derives from scale effects other than the division of labor. It is sometimes maintained that population growth can lead to per capita income growth, e.g. Simon (1977) and Boserup (1981). Such scale effects may clash with the economist's intuition of diminishing returns, but up to some point at least there are fixed costs and indivisibilities such as roads, schools, property-rights enforcement agencies, and so on, which can be deployed effectively only for large populations.

d) Growth that derives from increasing the stock of human knowledge, which includes technological progress proper, as well as changes in institutions. By technological progress, I mean any change in the application of information to the production process resulting either in the production of a given output with fewer resources (i.e., lower costs), or the production of better or new products. Again following Parker, I shall refer to this type of process as "Schumpeterian Growth." The choice of the words "application of information" is deliberate: much growth is derived from the employment of previously available information rather than the generation of new knowledge (Rosenberg (1982), p. 143). Traditionally, economists have distinguished between "invention" (the creation of new information) and its implementation, usually referred to as "innovation". Innovation is followed by "diffusion" in which more and more producers gain access to the new technique. In practice, these distinctions are not very helpful. During the implementation stages, inventions were usually improved, debugged, and modified in ways that qualify for the invention label. Diffusion, too, often required adaptation to local conditions and often implied further productivity gains due to learning by doing. Moreover, inventions that were not implemented remain little more than curiosa, of interest to intellectual but not economic historians. Some societies, such as the Hellenistic Mediterranean, were not especially technologically creative although they were capable of inventing. Their inventions for the most part remained amusing toys and had little or no economic impact. On the other hand, major gains in economic welfare were achieved due to exposure effects that occurred when previously disjointed civilizations came into contact with each other and learned to use each other's technologies. It is not really material, then, whether the information applied is entirely new even if a proper definition of what "new" exactly means here could be agreed upon. In what follows, the main focus is on technological creativity, defined as novel ways to apply knowledge so as to improve production techniques, a shift outward of the supply curve. Some societies have a remarkable record of technological creativity. Most have not.

The historical record of technological change displays an uneven and spasmodic character. Some brief spans in the history of a nation, like Britain between 1760 and 1800, are enormously rich in technological change. These peaks are followed by periods during which technical progress peters out. Why does that happen? Economists, sociologists, and historians have written extensively about this question, and they have found that its explanation is far from easy (Heertje (1983)). This essay is one more exploration in that direction. Most of what follows, however, is more of a historical description than an explanation. The reader interested in making a point about the history of technological change is offered a grab-bag of examples taken from the records ranging between 500 B.C. and 1914. The richness of technological history is such, however, that almost any point can be contradicted with a counterexample. Picking up empirical regularities in this massive amount of qualitative and often very uncertain information is hazardous and must be deferred (Mokyr, 1989a, 1989b).

The literature produced by modern economists on technological change is vast.[2] It has not, however, been very successful in explaining why some societies are technologically more creative than others. It is rarely informed by economic history, confining itself mostly to the post 1945 period. It is rarely informed by technological history. When technological historians such as A. P. Usher are cited, it is more for his interesting but speculative application of *gestalt* psychology to invention than for his enormous knowledge of how machines actually evolved over time (see for instance Thirtle and Ruttan (1987), pp. 2–5). Economists typically approach the explanation of technical change by considering the relation between demand and supply variables, research and development, and productivity growth. In so doing, they implicitly treat technology as a input, albeit one with peculiar features, that is produced and sold in the market for research and development. Such a market may or may not be a useful description of the post-1945 period (Jewkes, Sawers, and Stillerman (1969)). It is clear, however, that for an explanation of the diffusion of windpower in medieval Europe or the adoption of intensive husbandry in seventeenth century Britain, such a framework is inappropriate. Technological change throughout most of history can hardly be regarded as the consequence of an orderly process of Research and Development. It possessed few elements of planning and precise cost-benefit calculation. How, then, to explain it?

Once the economist ventures outside the safe realm of traditional microeconomics and agrees to consider extra-economic factors, he or she often discovers that events are hopelessly overdetermined. Theories of technical change based on geographic, political, religious, military, and scientific factors are typically easy to concoct and hard to reject. Many explanations make sense. But are they

[2] For recent surveys see Thirtle and Ruttan (1987); Baldwin and Scott (1987); Coombs, Saviotti and Walsh (1987); Wyatt (1986).

likely to be correct? This way of posing the question may not be very useful; perhaps it would be better to ask whether they are persuasive. Can we amass enough evidence to show that a particular theory is supported by facts in addition to logic? In what follows below I shall try to follow this kind of methodology.

By focussing on Schumpeterian growth I am not abandoning other forms of economic growth. Technological change unaccompanied by other forms of growth is rare. The four forms of growth aid and abet each other in many complex ways. For example, the embodiment hypothesis maintains that much technological change is contained in new capital goods. Thus in the absence of capital accumulation, technological change would be slow. Scale effects spurred by demographic growth lead to technological change, according to Simon (1977), because a large population implies a larger probability of a clever inventor being born. Schumpeterian growth in shipping leads to increased gains from trade by reducing transportation costs. The main focus here will be on technological change proper. The other forms of economic growth will be dealt with only insofar as they touch upon it directly. Technological change has a *macro* as well as a *micro* aspect. Even though the processes of invention and adoption are usually carried out by small units (individuals or enterprises), growth itself is an aggregate process. The economic historian is therefore directed to the macrofoundations of technological creativity, that is to say, what kind of environment makes individuals innovative, what kind of stimuli, incentives, hopes, and fears create an economy that encourages technological creativity?

It may turn out, as Heertje (1983, p. 46) has put it, that technological change cannot be explained. By that he means, I think, that standard economic theory faces a dilemma in dealing with technological creativity. It deals, after all, with rational choices subject to known constraints. Technological change involves an attack by an individual on a constraint everybody else takes as given. Yet most of the literature in the economics of technological change has viewed it as a process in which certain inputs (research and development) are transformed into an output called "productivity growth". As a result, the economics of technological change has been side-tracked by secondary questions such as its direction (factor saving biases) and the effect of cyclical variation in demand on the rate of patenting.

An essay in the economic history of technological change inevitably contains dates, names, and places. By its nature, the tale of technological creativity requires mentioning who first came up with an idea, when and where. Yet in the past decades, economic historians have not practiced this type of history. As David (1987) asks, does technology not simply accumulate continuously from the incremental, almost imperceptible, changes brought about by a large number of anonymous people? Some historians insist that almost all technological change consists of this "technological drift" (as Jones (1981)) has termed it, consisting mostly of invisible, small, incremental improvements (Rosenberg (1982), pp. 62–70). This incremental and evolutionary view of technological change, originally proposed by Gilfillan (1935) has been a good antidote to the naive "great man" interpretations of coffee-table books on the history of inventions. But it is possible to go too far in the other direction and discount major inventions too much. Some, like the printing press, the windmill, and the gravity-driven clock contradict the generality of the gradualist model of technical progress. There were, and probably always will be, large and discrete changes in technology which sweep the world off its feet and make it rush in to acquire and imitate the novelty. Modern research has shown, to be sure, that most cost-savings are achieved through small, invisible, cumulative improvements. But improvements in what? Virtually every invention ever made was followed by a learning process, during which the cost of the technique fell; but for them to fall, the technique had to come into existence first. An adult weighing 150 pounds has acquired about 95 percent of his weight since birth—does that mean that conception is not important?

In discussing the distinction between minor inventions whose cumulative impact is decisive in productivity growth, and major technical breakthroughs, the history of technology may benefit from drawing an analogy with the modern theory of evolution.[3] The biologist Richard Goldschmidt (1940) distinguished between micromutations, which are small changes in an existing species and which gradually alter its features, and macromutations which create new species altogether. There is considerable dispute whether the Goldschmidt scheme is appropriate for modern evolutionary biol-

[3] For a more detailed exposition of this comparison, see Mokyr (1989a, 1989b).

ogy (Eldredge, 1985, but see Dawkins, 1986), but in any event the distinction between the two is useful for our purposes. In analogy with Goldschmidt's distinction, define microinventions as the small, incremental steps which improve, adapt, and streamline existing techniques already in use, reducing costs, improving form and function, increasing durability and reducing energy and raw material requirements. Macroinventions are those inventions in which a radical new idea, without clear precedent, emerges more or less *ab nihilo*. Whereas in terms of sheer numbers microinventions are far more frequent and from an economic point of view they account for most gains in productivity, the importance of macroinventions in technological history is crucial.

The essential feature of technological progress is that the macroinventions and microinventions are not substitutes but complements. Without subsequent microinventions, most macroinventions would end up as curiosa in a museum or in someone's sketchbook. Indeed, in some historical cases the person who came up with the improvement receives more credit than the inventor responsible for the original breakthrough, as was the case with James Watt and his improvements to the Newcomen atmospheric engine. But without novel and radical departures, the continuous process of improving and refining existing techniques would run into diminishing returns and eventually peter out. Microinventions are more or less understandable with the help of standard economic concepts. They result from search and inventive effort, and respond to prices and incentives. Learning by doing and learning by using increase economic efficiency and can be related directly to economic variables, capturing myriads of infinitesimal inventions. Macroinventions, on the other hand, do not seem to obey obvious laws, do not necessarily respond to incentives, and defy most attempts to relate them to exogenous economic variables. Many of them resulted from strokes of genius, luck, or fortunate misunderstandings. Technological history, therefore retains an unexplained component which makes a purely economically oriented explanation difficult to maintain. In other words, luck and inspiration mattered, and thus individuals made a difference.

For that reason, a survey such as this one has to be concerned with a succession of unique actions and discrete historical events. The notion that if an invention had not been made by X it would have been made by Y, inferred typically from the large number of simultaneous discoveries, is a misleading overgeneralization. It may be true for the telephone, the incandescent lightbulb and other examples; it is false for scores of other important inventions. If there is any area in which a deterministic view that outcomes are shaped inexorably by forces stronger than individuals—be they supply and demand or the class struggle—is oversimplified, it is in the economic history of technology. Asking whether the breakthroughs are more important that the marginal improvements is like asking whether generals or privates win a battle; the processes are complementary. Yet it is useful to organize the narrative around the discrete event. Just as in military history we employ shorthand such as "Napoleon defeated the Prussians at Jena in 1806," we may say that so and so's invention occurred at this and that time. We do not imply that inventors are credited with all productivity gains from the invention any more than that Napoleon defeated an entire Prussian army single handedly.

One final point: history provides us with relatively few examples of societies that were technically progressive. Our own world is exceptional though not unique in this regard. Consequently, this study will necessarily draw from a small sample and be heavily biased toward Western economic history. I shall leave out the prehistoric and very early eras, and limit the discussion of the post-1914 period to a minimum. Between about 500 B.C. and 1914 there is a record so stupefyingly abundant in facts and evidence, that I can only scratch the surface of a deep and rich seam.

2. CLASSICAL ANTIQUITY

Invention occurs at the level of the individual. It is a game against nature, rather than against other players, what Von Neumann and Morgenstern call a "Crusoe game."[4] Yet the adoption and diffusion of technological change involve social interaction, and the kind of society in which the inventor lives decides whether the invention will be taken up by others. Many societies on record did not provide

[4] As one historian of technology has noted, "a group can conceive nothing which is not first conceived by a person" (White (1968, p. 105)).

the right environment in which inventions would spread at a pace strong enough to affect output and thus living standards perceptibly. The consensus on classical civilization (Greek, Hellenistic, and Roman) is that these societies were not very successful from a technological point of view (Finley (1965, 1973); Lee (1973); Hodges (1970)). As some critics have pointed out, such a judgment is overly harsh (K. D. White (1984)). First, classical civilization *did* make some significant breakthroughs, and their extent is likely to be underrated by historians because of the scant literary and archaeological evidence remaining. Second, the notion that science could be applied to attain concrete objectives rather than admired for its own sake was certainly present, and particularly developed amongst the Hellenistic mechanicians. Third, as Finley (1973, p. 147) emphasizes, in some sense the judgment imposes our own value system on a society that had no interest in growth. "As long as an acceptable life-style could be maintained, however that was defined, other values held stage." In the areas that mattered most to *them,* the Greeks and Romans achieved huge successes. Nonetheless, the judgment reflects our intuitive disappointment with a civilization that has celebrated such triumphs in literature, science, mathematics, medicine, and political organization.[5] The greatest achievements of classical technology were in those aspects of technology that were non-physical in nature: coinage, alphabetization, stenography, and geometry were part of the information processing sphere rather than the physical production sphere of the economy.

What technological progress there was in the classical world served the public, rather than the private sector. Because most agriculture, manufacturing, and services were carried out by the private sector, achievements there were few and slow. The major areas of achievement for which the Greeks and Romans became famous were in civil and hydraulic engineering and architecture. Water conduits, built both for the supply of fresh water and

[5] By comparison, the early iron age (1100 B.C. to 500 B.C.) saw, in addition to the development of the use of iron, the development of most carpenters' tools (lathes, saws, pegs), shears, scythes, axes, picks, and shovels. In this age the rotary quern (used to grind flour and ores) was invented, and the construction of ships and wagons vastly improved.

drainage, date from early classical Greece.[6] The Roman Empire, which commanded enormous resources, brought public-good engineering to amazing heights, although most of their engineering feats, including roads and aqueducts, utilized existing technology. The supply of Rome with water was begun by Appius Claudius in 312 B.C. and the system reached unprecedented complexity by the first and second centuries A.D.[7] Sewage and garbage disposal were also highly developed. A system of central heating was developed for use in both homes and bath-houses. In spite of the urban bias in Roman technology, Hodges (1970, p. 197) concludes that "the Roman city was more interesting for its scale than for the novelty of its design."

Equally important in Roman engineering was the infrastructure of land transport: the roads and bridges built by the Romans are justly considered as one of their proudest achievements. Much of their success in these works was due to the Romans' discovery of cement masonry, which one historian of technology has called "the only great discovery that can be ascribed to the Romans."[8] In bridges and aqueducts, the Romans used the revolutionary technique of weight-supporting arches or pillars made of cement. Some of their aqueducts such as the famous Pont du Gard near Nîmes have been preserved. Others, such as the wooden trestle bridge that Julius Caesar's troops built on the Rhine in ten days (in 55 B.C.) are known to us only from documents. A further area where technical ingenuity improved efficiency in the public sector was in the construction of war machines. Although military technology will not

[6] The first such large project, the Samos aqueduct, was built by Eupalinus of Megara, around 600 B.C.. It brought water from a lake to the town along a tunnel about a mile in length and eight feet in diameter.

[7] By the time of the water superintendent Frontinus, writing between 97 and 104 A.D., many houses in Rome had hot and cold running water. The use of lead pipes was an ingenious solution to the short supply of construction material, but may have had severe side effects on the population's health. Roman writers, interestingly enough, were aware of the problems of lead poisoning, but their warning was not heeded, and it was not until the time of Benjamin Franklin that lead poisoning was again identified as a hazard (K. D. White (1984), pp. 164–65).

[8] See Forbes (1985b, p. 73). This credit is not entirely deserved, since cement was used in Asia Minor before Rome. But the Romans vastly improved its use, and employed such quality control that they may well be thought of as the originators of the technique. Specific to Roman construction was the use of water-proof cement, known today as *pozzolana*, made from the volcanic dust found near Naples.

be of major concern here, it is worth noting that Greek and Roman military technology provides one of the few areas of a successful collaboration between scientists and technology.[9] Oddly enough, the Romans did not improve much on Greek and Hellenistic military machines, although they made extensive use of them and built bigger and more powerful versions.

In what we would today call "machines," the contribution of the classical world, especially Hellenistic civilization, was confined mostly to gadgets and clever machines that were admired for their own sake but rarely put to a useful purpose. Many of these ideas subsequently lay dormant for millennia. Perhaps the most brilliant inventor and engineer of antiquity whose works have come down to us is Hero of Alexandria, whose dates are somewhere in the late first century A.D. (Landels (1978), p. 201). Among the devices credited to Hero are the *aeolipile,* a working steam engine used to open temple doors, a coin-operated vending machine (for holy water in the temple), and the *dioptra,* an instrument used in surveying and construction, combining the theodolite (measuring angles) with the level. Most of Hero's inventions served at best recreational purposes. The same can be said about Ctesibius, whose dates are uncertain, reportedly the inventor of the hydraulic organ, the water clock, and the force pump.

A recent discovery of a product of the technical genius of Hellenism is the famous Antikythera mechanism, found as cargo on a sunken ship near Crete. It is a geared astronomical computing machine of astonishing complexity, built in the first century B.C. Derek de Solla Price, who reconstructed the mechanism, admonished historians to "completely rethink our attitudes toward ancient Greek technology. Men who could have built this could have built almost any mechanical device they wanted to" (Price (1975), p. 48). This statement seems to push revisionism too far. What the mechanism proves is that Hellenistic civilization had mastered the use of gears and applied geometry beyond what was believed possible, and that astrolabes (invented in the second century B.C.)

[9] The story of Archimedes helping to build war machines in a futile attempt to defend his native Syracuse against Roman armies, is well known (Hacker (1968)). Other engineers who made major contributions to the construction of catapults are Ctesibius (third century B.C.) and Philo of Byzantium (about 180 B.C.).

were mechanically sophisticated. The Antikythera machine demonstrates that classical civilization had the ability to build sophisticated astronomical instruments and could construct far more complex geared mechanisms than has hitherto been thought. Yet the mechanism was, after all, a gadget used to reproduce the motions of the moon, the sun, and the planets for scientific and probably astrological purposes. It served, as far as we can tell, no direct economic purpose and in the view of one classical scholar the special branch of machine design that made the Antikythera machine was not representative of the mainstream of invention of the period, nor did it have much impact on it (Brumbaugh (1966), p. 98).[10] What it shows is that classical civilization had the intellectual potential to create complicated technical devices. The question remains why so little of this potential was realized and translated into economic progress.

One of the activities in which Hellenistic and Roman civilization made lasting contributions was in water lifting and pumping. Pumps of various designs were in wide use in antiquity, including irrigation, mine drainage, fire fighting, and bilging water from ships. Force pumps, which were known and used by Romans, had the disadvantage that they had to be immersed in water, thus making them more difficult to operate and install. However, the suction pump, the obvious complement to the force pump, was not invented until the fifteenth century (Oleson (1984); Landels (1978), ch. 3). Water lifting devices led to some advances in mechanics, such as the development of power transmission (gears, cams, and chains). Yet identifiable positive externalities from water lifting to other industries are fairly rare, and some important mechanical insights such as the crank and the flywheel were missed.

In the private sector, such as agriculture, textiles, and the use of power and materials, the progress achieved between 500 B.C. and 500 A.D. was far from overwhelming. New ideas were not altogether absent, but their diffusion and application were sporadic and slow. From a purely economic point of view the realization of the principle of the lever, attributed to Archimedes, was the most

[10] Derek Price has argued that astronomical instruments eventually led to the incidental invention of clocks and permitted the measurement of time as a by-product. This theory is effectively dismissed by Landes (1983, pp. 54–57).

important innovation. The lever was combined with the principle of the helix to produce the screw and used for gearing, attaching, and pressing. What connects these techniques is the realization by ancient engineers that a small force operating over a distance can exert larger and larger force as the distance gets progressively smaller. The compound pulley, which allowed cranes to lift large weights, was another example of this. What part of these innovations was directly owed to Archimedes is not clear; it is likely that here, as in many other instance, science followed technology rather than vice versa.[11]

In metallurgy and mining, the main advances were the use of scoopwheels and Archimedes screws. The Greeks made some improvements in separating extracted minerals from waste, but by and large there were few changes of substance. Greek and Roman iron making was slow, and produced an uneven, and thus quite poor, product. The ore was heated in a "bloomery process" which burnt the carbon and left a slag of contaminating minerals which had to be removed to produce what we could call today wrought (i.e., carbon-free) iron. The binding constraint on metallurgical technology was the inability of ancient blacksmiths to produce iron at temperatures high enough for casting. The spongy and pasty blooms produced by ancient blacksmiths needed frequent hammering and further heating in order to yield a metal that was at best serviceable. The best quality steel, "seric iron" (or wootz steel) was imported from India, although the bloomeries of the West were capable of producing steel of low quality (Barraclough, Vol. I, p. 19). In the fourth century A.D. bellows were probably in use in furnaces, but cast iron was unknown because sufficiently high temperatures could not be generated. Here the classical and early medieval world lagged behind China, where the art of casting iron had been known since the third century B.C.. As far as we can determine, the Greeks and Romans made little progress in metallurgy, in spite of their heavy use of iron. The best that can be said is that the effect of the Roman Empire was the dissemination of best-practice techniques, and possibly somewhat larger furnaces, though the evidence is weak and tends to be mostly for Eastern

[11] We can find antecedents for the use of pulleys in earlier periods. It seems that in the eighth century B.C. the Assyrians already possessed a pulley type of device.

Europe and Britain, hardly typical of the Mediterranean world ruled by Rome (Tylecote (1976), p. 53).

The same is true for shipping. Shipping was crucial to the prosperity of the economies of the Mediterranean: prosperity in classical times was founded primarily on commerce, that is, on the gains from "Smithian growth," made possible through a complex network of interregional trade supported by colonization and later by Roman laws and political control. Classical cargo ships were equipped with a single square mainsail, sometimes supplemented with smaller topsails. The myth that ships in classical times could not manoeuvre against the wind has by now been satisfactorily laid to rest (Casson, (1971, pp. 273–4); K. D. White (1984, p. 144)). Even square-rigged vessels could beat against the wind, through they were designed primarily to sail with the wind and could probably do no better than "one point into the wind," that is, at an angle of 79° against the wind. The best way to manoeuvre against the wind is to use fore-and-aft rigging (in which the sail is set parallel to the keel). Casson has argued that classical civilization knew fore-and-aft rigging in at least three forms: the sprit sail, the triangular lateen and the quadrilateral lateen. The evidence for the existence of these sails is persuasive, but it is also clear that large merchant ships could not use these riggings and that the primitive square sails remained prevalent; for many years historians believed that square rigs were the only rigging the Greeks and Romans knew. The reasons proposed for the delay of widespread fore-and-aft-rigging include the lack of suitable trees for the larger masts, the slower speed of fore-and-aft rigged ships with the wind, and the difficulty of shifting the sail during tacking (the zig-zag motion that allows sailing ships to move against the wind). Steering by the means of oars was difficult, although the Romans achieved some improvement here using a pivot device that made manoeuvering a ship a little easier.

Agricultural technology made mostly local advances. The tools and techniques used in Roman times show great variety, adapting the basic primitive techniques of Mediterranean farming to local conditions. Labor saving inventions were rare, and K. D. White's statement that "technical development in Roman farming was common while innovation remains rare," summarizes the historical evidence (K. D. White (1984), p. 58). We know that the Romans tried to breed cattle for the sake of fertilizer, but the eternal problem of how to feed the livestock was not solved satisfactorily. The Roman successes in hydraulic engineering found a few applications in drainage and irrigation, but their marginal contributions in these regards pale compared with the huge hydraulic irrigation works of Egypt and Mesopotamia two millennia before them. Whatever change occurred in agricultural implements came from outside the Mediterranean perimeter: the Gauls improved harvesting equipment, and there are a few famous but controversial accounts of primitive harvesting and threshing machines, but there is no evidence that these devices were used widely or had a major impact on productivity.[12]

One of the most famous inventions made during the Roman Empire was the waterwheel. The Roman engineer Vitruvius, who lived in the second half of the first century B.C., describes one in which gears were used to transmit power from the turning wheel to the grindstones. Even earlier references are known, but they do not describe the machinery in any detail.[13] As in the case of fore-and-aft rigging, classical society could create a novel technique, but was unable or unwilling to take it to its logical conclusion and to extract anything approximating its maximum economic rent from it. Thus, until the fourth or fifth century A.D., waterwheels were apparently used only for flour milling, and there too their use was not widespread. Perhaps the reason was that Vitruvian mills were undershot mills, which were not very efficient, though generally cheaper to build and more suitable to the Mediterranean topog-

[12] The harvesting machine developed in Gaul known as *vallus* was essentially a reaping oxcart with teeth-like cutting devices in front to cut the grain. It was last mentioned by Palladius in the fifth century, and its subsequent disappearance suggests that it may not have worked very well. The classical period also saw the invention of the Roman sickle and the British long scythe (K. D. White, (1984), pp. 49–50). Yet roman scythes were so rare that their dating was long doubted (Lynn White (1972, p. 149).

[13] Antipater of Thessalonika wrote in 85 B.C. the following epigram:

Cease from grinding, ye women who toil at the mill
sleep late even if the crowing cocks announce the dawn
For Demeter has ordered the (water) nymphs
to perform the work of your hands
and they, leaping down on the top of the wheel,
turn its axle which with its revolving spokes,
turns the heavy concave Nysarian millstone.

(Cited in Usher (1954), p. 165).

raphy. More efficient overshot wheels, using gravity rather than the impulse of the movement of water appear in the West possibly as early as the third century A.D. (Reynolds (1983), p. 26), and can be dated with certainty to Athens in the middle of the fifth century (Blaine (1976), p. 166).

In assessing classical technology, it is important not to overemphasize the lack of achievement. Our sources are biased, because many tools and devices, made out of wood or other organic material, have not survived. Most classic writers did not bother with technological matters, which may be a significant fact by itself.[14] Archaeological evidence has demonstrated recently that waterwheels were more widespread than literary sources had suggested (Wikander (1985)). Derek Price has argued that what we do for ancient civilization is tantamount to reconstructing modern society on the basis of a few buildings and paintings. "The technology was there, and it has just not survived like the great marble buildings ... and the constantly recopied literary works of high culture" (Price (1975), p. 48). Yet as Finley (1965, p. 29) points out, there is a reverse bias as well. The first appearance of an invention does not correspond necessarily with a major effect on productivity, which requires widespread adoption. When all is said and done, it remains something of an enigma why a commercial and sophisticated economy, heavily dependent on transportation and animate power, in which handicrafts and food processing for a large urban population were widespread, failed to reach some rather obvious solutions to technical problems that came to the fore in the first centuries of what we now call the Middle Ages.

3. THE MIDDLE AGES

It is customary to divide the Middle Ages into an early period, until about 1150, and a late period between 1150 and about 1500. Early medieval Europe, which is still sometimes referred to as a "dark"

[14] Thus, Vitruvius, in his *De Architectura*, remarks that "there are innumerable mechanical devices about which it does not seem useful to enlarge because they are too handy in our daily use."

age, managed to break through a number of technological barriers that had held the Romans back. The achievements of early medieval Europe are all the more amazing because many of the ingredients that are usually thought of as essential to technological progress were absent. Particularly during the first three centuries, between 500 and 800 A.D., the economic and cultural environment in Europe was primitive compared to the classical period. Literacy had become rare, and the upper classes devoted themselves to the subtle art of hacking each other to pieces with even greater dedication than the Romans. Commerce and communications, both short- and long-distance, declined to a trickle. The proud roads, bridges, aqueducts, ports, cities, and villas of the Roman Empire fell largely in disrepair. Law enforcement and the security of life and property became more and more precarious as predators from afar and nearby descended upon Europe with a violence and frequency that Roman citizens had not known. And yet toward the end of the dark ages, in the eighth and ninth centuries, this society began to show the first signs of what eventually became an astounding technological creativity. Not the amusing toys of Alexandria's engineers or the war engines of Archimedes, but useful tools and ideas that reduced the daily toil and increased the material comfort of the masses, even when population began to expand after 900 A.D. Compare the progress achieved in technology in the seven centuries between 300 B.C. and 400 A.D. with that of the seven centuries between 700 and 1400 A.D., and prejudice against the Middle Ages dissipates rapidly.

Medieval technology drew from three sources: its own original creativity, antiquity, and technical ideas in use in Islamic and Asian societies. Medieval engineers did not seem to care much where ideas came from as long as they worked. In the majority of cases we do not know where the ideas came from. At times what seems like the adoption of a *known* technique was in fact re-invented independently. It is also possible that what seems like an original idea in society *A* was in fact borrowed from *B*, but no record of its existence in *B* has survived. In any event, diffusion of new technology was usually slow, and the old techniques often stubbornly survived and coexisted with the new for decades and even centuries.

In terms of their direct contribution to aggregate output, changes in agricultural technology were particularly important, as the vast bulk of the population was engaged in farming. Yet changes here were especially slow. Agricultural technology differs from manufacturing, transportation, or information technology in that it tends to be highly site-specific. Different crops have different requirements, and the same crop will use different inputs and technology depending on elevation, rainfall, soil type, and so on. As a result, improvements have to be modified and adapted infinitely, and thus a significant part of the development cost is imposed on the user of an innovation, and the additional experimentation slows down the process. The transformation of agriculture that started in the early Middle Ages took many centuries to complete, but eventually it shaped European history.

The essential elements of the agricultural evolution were the introduction of the heavy plow and the creation of the three-field system. It has taken the combined geniuses of Marc Bloch (1966) and Lynn White (1962) to make historians fully realize the importance of the heavy plow or *carruca*. The ancient plow, used in the Mediterranean economies, scratched the soil with a wooden or iron cutting point or "stock," which cut and pulverized the soil, preventing the evaporation of moisture and bringing subsoil minerals to the surface by capillary action. This *aratrum* was ill-suited to the heavy and moist clay soils of the plains north of the Alps. The heavy plow, in its ultimate form, moved on wheels, and had a coulter that cut the soil vertically, a flat plowshare that cut horizontally, and a mouldboard that turned the cut sods aside to create a deep furrow. The heavy plow made possible the cultivation of huge tracts of fertile land that in Roman times were either uncultivated or exploited by a primitive slash-and-burn technique. Although some Roman plows had wheels, the complete heavy plow did not make its appearance before the sixth century (White (1962), p. 53).[15] The heavy plow created the peculiarly long and narrow strips that characterized European open fields. But its impact was especially momentous because it required a team of oxen to pull

[15] The exact timing of the introduction of the complete wheeled plow into the plains of Northern and Western Europe is a matter of some dispute. The evidence supporting White's claims has been disputed by other scholars, especially regarding the mouldboard. See for example Wailes (1972).

it.[16] Few peasants could afford to own such an expensive capital good, and in part in an attempt to solve the fixed cost problem, medieval society developed a semi-cooperative organization sometimes referred to as the manorial system.

The dependency of plowing on draft animals created the technical problem with which European agriculture grappled for many centuries: how to feed its animals. The solution found in the early Middle Ages combined three elements, though not all three of them were necessarily present together. First, under the new three-field system of crop rotation that spread slowly through Europe in the early Middle Ages, one third of the arable land was left fallow. The animals were let to graze on the fallow land, thus feeding themselves while at the same time fertilizing the soil with their droppings. Second, the fields under crops were opened up to stubble grazing after the harvest, a custom know as "the right of common stock" or *vaine pâture*. Third, the village usually had a "common" field on which animals grazed. The open field system was not a technical invention *strictu sensu* as much as a brilliant organizational solution to a technical problem by combining private and public property rights in an ingenious fashion. The introduction of the open field system permitted the expansion of the cultivation of additional crops beside the main staples such as wheat and rye. The second field was often used to grow oats (an ideal horse fodder), barley (an industrial raw material as well as a human food), and legumes (an important dietary supplement). None of these crops was novel, but their expansion depended on the replacement of the biennial with a triennial rotation.

A second area in which early medieval Europe was successful was energy utilization. Energy takes two forms: kinetic energy and thermal energy. Kinetic energy could be derived from two sources: animate power (including human muscles), which transmitted the energy of the sun through living bodies, and non-animate power that used solar power directly through water- or wind motion. Wind power had been used in sailing ships, but had not been harnessed in the West in other ways until the first windmills were built there in the twelfth century. In waterpower, on the other

[16] Pliny's Roman wheeled plow was pulled by eight oxen. In the Middle Ages the more normal number was four to six.

hand, the early Middle Ages saw radical improvements. In the early Middle Ages better and bigger waterwheels spread through Europe. Medieval Europe produced not only the more efficient overshot wheel, but also adapted and improved the gearing of both horizontal and vertical waterwheels and thus made it possible to use wheels on both rapidly flowing rivers such as the Tiber and slower flowing streams. Medieval engineers made much progress in the construction of dams which allowed a more controlled usage of water power through storage and diverted streams to mill races leading to convenient sites. They applied cams and later cranks to convert the circular motion created by waterwheels into the reciprocating motion needed for hammering, fulling, and crushing. The cam had been known in antiquity but was apparently not combined with the waterwheel, while the crank was a medieval invention. The result was that the waterwheel turned from an occasional device grinding flour into a ubiquitous source of energy operating on flowing water of every type. By about 1100, waterpower was used to drive fulling mills, breweries (to prepare the beer mash), trip hammers, bellows, bark-crushers, hemp treatment mills, cutlery-grinders, wire drawers, and sawmills. The Domesday book of 1086, the first "census" of an entire country, listed 5,624 watermills in England south of the Severn or roughly one for every fifty households. Unlike their ancestors in Roman times, medieval men and women were surrounded by water-driven machines that did much of the more arduous work for them.[17] Medieval Europe may not have invented the waterwheel, but it succeeded in its use far beyond anything in earlier times. As Lynn White has remarked, it was perhaps the first society to build an economy on non-human power instead of on the backs of slaves and coolies.

In animate energy, too, there was progress. The horse had been used for millennia before the Middle Ages by man, in war and peace. In the centuries subsequent to the fall of Rome, three important improvements vastly increased the efficiency of the use of

[17] In the eleventh century, there were even tidal mills in Europe (around Venice, in the south of England, and on the west coast of France), the first use in history of power from non-solar origins (Minchinton (1979)). Although this source of power never became important, it is symptomatic of the thirst of medieval people for inanimate energy wherever possible. See Derry and Williams (1960), pp. 252–53, and Gimpel (1976), p. 23.

horses. The first was the invention of the nailed horseshoe, around the ninth century. The horseshoe was especially useful in the moist soils prevalent north of the Alps and for the heavier horses used in the latter Middle Ages. A hipposandal for horses had been used in antiquity, but from every point of view the horseshoe represented an improvement. The second was the invention of the stirrup, whose value was mostly military, but of course it also benefitted civilian riders.[18] The third and most important innovation regarding horses was the modern horse collar. In the beginning of the twentieth century a French cavalry officer, Richard Lefebvre des Noëttes, spent his retirement years comparing the use of horses in antiquity and the Middle Ages. He discovered that the Greeks and the Romans used a yoke harness in which two straps were wound around the belly and neck of the horse. The neck strap simultaneously pressed on the animal's jugular vein and cut off the windpipe as soon as it began to exert pressure. Experiments indicated to Lefebvre des Noëttes that a horse strapped this way lost about 80 percent of its efficiency. The modern harness, originally developed in China, is rigid and rests on the shoulders. It was more or less in place in the ninth century. Consequently, horses acquired major economic importance in agriculture and in the hauling of wagons. It is amazing that for hundreds of years the sophisticated civilizations of the Mediterranean made so egregious an error in one of the most basic applications of animate power, but the historical record is quite clear here.[19] Medieval society, by way of contrast, did not tolerate such an easily corrected waste of valuable energy.

Many other inventions date from the early Middle Ages, although timing and placing them with any accuracy is impossible. In ship-design, the *cog* was gradually developed from primitive Celtic

[18] In a famous but controversial argument, Lynn White (1962) has maintained that the stirrup was directly responsible for the emergence of "feudalism" which, if true, is a remarkable example of technological determinism. The stirrup established for centuries the unassailable superiority of the horseman over the foot soldier, leading to the need to equip and arm large number of knights. Because horses and iron were both scarce, the entire economic system had to be geared toward paying for these armies. This theory has come under criticism, and it is likely that feudal society had other roots as well. For our purposes it is important primarily as a demonstration of the determination and ability of medieval persons to utilize energy efficiently.

(*Footnote 19 overleaf.*)

boats. Rather tubby-looking, slow, and awkward to manoeuvre, it proved eminently seaworthy, capable of carrying large cargoes at low cost, and easy to defend against pirates. In medieval times, European society learned gradually to protect itself from cold temperatures, especially as winters grew harsher after the thirteenth century. Changes in construction techniques, such as the use of shingles and better insulation through plaster, the use of fossil fuels, and above all the development of chimneys, testify to the ability of medieval society to struggle against the harsh winters of Northern Europe (Dresbeck (1976)). The chimney also facilitated home cooking, thus allowing people to eat warm foods. Soap, while known in the late Roman Empire, became widespread in Europe around 800 A.D. and was improved upon in the tenth and eleventh centuries in both the Christian and Moslem regions of the Mediterranean. Cakes of hard soap first appear in the twelfth century. Butter, distilled strong liquors, skis, the use of hops in brewing, improved window glass, and crank-driven grindstones originated in western Europe in the early Middle Ages.

In the early Middle Ages, the cultural and technological center of gravity of Europe remained to a large extent in the Mediterranean region. Spain, North Africa, and the Middle East were ruled by Islam. Between the eighth and the twelfth centuries, Moslem society was more sophisticated and better educated than Western Europe. Medieval Islam was also a highly mobile society, in which travelling was a normal activity, enjoyed by rich and poor alike, and travellers were eager to learn from other societies, past and present. The culture and technology it created constituted a synthesis between Hellenistic and Roman elements, adorned with ideas from Central Asia, India, Africa, and even China. Early Moslem society

[19] An attempt to explain this failure of classical technology is provided by Landels (1978, pp. 176–77), but the lameness of his argument is self-evident. He notes that the chariots pulled by horses in antiquity were small and light, so that no great effort on their part was required. Here he places his own horses before the cart: the chariots were small *because* there was no way to pull them efficiently by horses. It is true, however, that the use of horses as draught animals also depended on the ability to feed the horses adequately and to breed horses large enough to provide enough power (Barclay (1980), p. 109). It is also important not to claim too much for inventions, even significant ones. Lefebvre des Noëttes claimed that the horsecollar was the chief cause of the disappearance of slavery, a claim that is without much merit.

collected, compiled, and catalogued knowledge avidly. It was a literate society beyond the wildest dreams of Europe: not only the rulers, but all important mosques had large libraries. The Moslems knew more about the different parts of the known world than any other civilization between 700 and 1200. The ability of Islam to preserve, adapt, and develop techniques borrowed from others is a lasting testimony to their creativity. Yet when it came to original creativity and invention, the achievements of Islam appear strangely truncated. The Moslems were enthusiastic collectors, but they offered little in the way of interpretation or theory; unlike the West they were not driven, apparently, "by a strange urge to peer beneath the surface of things and see how they worked" (Watson (1983), pp. 94, 146). Moreover, the technological creativity of Islam civilization for some reason ran out of steam. At some time around the twelfth century, Islam lost its momentum, and technological supremacy eventually passed to their Christian enemies north of the Pyrenees, who were equally capable at borrowing technology from others and turned out to be better at creating their own.

Still, in the centuries between 750 and 1100, Islam had some impressive technological achievements to its credit, in addition to being a more tolerant and cultured society (Singer (1957), pp. 755–61). Moslems, together with the Byzantines, are associated with the development of the "lateen" (i.e., Mediterranean) sail, though earlier versions of it were known to the Romans. The idea of the lateen sail was revived by the sailors on the Mediterranean. By the ninth century the problems of shifting the sail spar over the mast during tacking, which had prevented the Romans from building larger ships using fore-and-aft riggings, had been solved, and lateen sails were adapted to larger merchant ships. With a lateen sail a ship could sail 60 or 65 degrees off the wind, which was "very much better than the performance of square-rigged ships" (Parry (1974, p. 13)). By the eleventh century these improved sails were used throughout southern Europe and the Middle East by both Moslems and Christians. In power technology, the Moslems were the first to use a tidal mill in Basra, around 1000 A.D. (Minchinton (1979)), and windmills are mentioned in Moslem sources from the ninth century on (Al Hassan and Hill (1986), pp. 54–55). Water power was used in sugar mills and saw mills; yet the overall reliance of Moslem society on inanimate energy appears much smaller than in Europe (Hill (1984b), pp. 169–172).

The Moslems were also responsible for the introduction of paper into the occidental world. The need for a substitute for the expensive parchments used by medieval scribes was evident, as parchment was made of dried calf or kid skins. Paper was invented by the Chinese before 100 A.D. and was widely used in the Orient. According to traditional accounts, when the Arabs conquered the city of Samarkand in 753, they learned the secret from captured Chinese workmen. In about 793, the first paper factory was set up in Baghdad, and by 1000 the entire Moslem world was enjoying bound books as well as wrapping paper and paper napkins. The penetration of paper into Northern and Western Europe was slow. Once it was introduced, in the thirteenth century, the West characteristically immediately used water power to process the pulp.

In textile production, the Moslem world made substantial advances in fabric quality. Here etymological evidence is suggestive: many names for European fabrics betray their Moslem origins. Damasks, a fine silk-like table linen, came from Damascus; mousselins or muslins, a fine cotton fabric, from Mosul (in Iraq); fustians, fabrics made of a linen warp with a cotton weft were named after Fustat (a Cairo suburb). Many of these fabrics may have originated from Persia, the cradle of many textile technologies, and moved slowly westward in the tenth century. Cotton production was introduced by the Moslems into the Middle East and from there into Spain and Sicily. The cotton variety they introduced into the Mediterranean was capable of maturing in a harsher and colder climate. By the end of the Middle Ages, in the part of the Old World south of the fortieth parallel (a line between Madrid and Ankara), cotton had become the main raw material of which clothes and other fabrics were made. In leather production, the Moslem economies were famous for producing high-quality goods. Morocco and Cordoba in Spain were centers of this industry whose names were immortalized in Western Languages.

Moslem chemical technology was the area in which their contribution was the most original. Al Jabir and Al Razi (known in the West as Geber and Rhazes) wrote books which for centuries were the recognized standard works in the field. Much work in chemistry was aimed at the useless search for gold and had a mystical element to it, but by the eleventh century Avicenna or Ibn Sina doubted that gold could be made at all. The Moslems invented alkalis, and much improved the quality of glass and ceramic products. They produced "naphtha," a flammable petroleum derivative akin to kerosene, and their perfume and acid industries were far superior to anything known before.[20]

In mechanical engineering from water mills to clocks, the Moslems were for centuries far ahead of the West. The Banu Musa brothers, writing in Baghdad in around 850 A.D. summarized and extended Hellenistic mechanical engineering (Hill (1977)). Al Jazari's *Book of Knowledge of Ingenious Mechanical Devices,* written in the early thirteenth century in Diyarbakir (Asia Minor), has been dubbed "the most remarkable engineering document to have survived from . . . pre-Renaissance times" (Hill (1984b), p. 128). In metallurgy too, Moslem technology thrived for a while. Toledo and Damascus produced the best steel swords in the Western world though the material itself came in part from India. The Arabs also learned to grow and refine sugar and to make confectionery from it. Egypt especially became famous for its sugar products, though they remained a luxury good until the seventeenth century. Many Asian crops that improved the Western diet were transmitted through the Moslem world; cereals such as sorghum and rice, and the hard durum wheat of which pasta is made; fruits such as oranges, bananas, cantaloupes, and watermelons; and vegetables such as asparagus, artichokes, spinach, and eggplants (Watson (1983)).

Indeed, agricultural progress, in the Moslem world between about 700 and 1100 A.D. has been such that the word "agricultural revolution" has been used. The adoption of new crops required in many cases irrigation, and the Moslems were masters in the utilization and modification of ancient hydraulic technology for use in the cultivation of these crops, most of them unknown in the Mediterranean before the advent of Islam. From Persia to the Pyrenees, there is evidence for a proliferation of norias, Archimedes screws, *shadufs* (a pole with a pail on one end and a counterweight on the other), and so on. For the most part, the

[20] There is some dispute as to whether Moslem scientists discovered alcohol distillation (Hassan and Hill (1986), p. 141; but see Lynn White, 1978, pp. 7, 115).

Moslems used techniques known in the classical world: their contribution was "less in the invention of new devices than in the application on a much wider scale of devices which in pre-Islamic times had been used only over limited areas" (Watson (1983), p. 108). From that point of view, their achievement was comparable to that of the West, which did the same for waterpower and heavy agricultural implements.

Assessing the technological achievements of the Moslem world is difficult. Not all Moslems were Arabs, and the "Moslem world" contained at different times large numbers of Christians and Jews. Persia and Spain, at the geographical extremes of the Moslem world, seem on the whole to have been more creative than Mesopotamia and Egypt, although the information is lacking to make such judgments with certainty. On the whole, the Moslem world made several important original contribitions to technology and served as a conduit of ancient and alien techniques to Europe. The verdict of technological historians on medieval Islam is, however, harsh. Wiet (1969, p. 371) writes flatly that it is difficult to discover anything that could be called an enrichment of acquired knowledge. Yet technological creativity as an economic phenomenon does not require the creation of totally new knowledge: borrowing, extending, and adapting will shift supply curves to the right just as well. What is striking about Moslem technology is that most Moslem societies lacked the stamina to keep improving upon a borrowed and imperfect technology, and subsequently lost momentum after 1150. The West, initially far behind, was equally willing to learn from other cultures, but never lost its capacity to improve and refine others' ideas, applying them in new combinations, adapting them to novel usages, and eventually surpassing the original idea to the point where the original inventing society had to borrow its own ideas back, often unrecognizably altered and improved. The reasons for the slowdown in Moslem technological creativity are unclear. The destruction of the eastern part of the Moslem world by Mongol invasions was a dreadful blow; yet the Moslem loss of Spain and the decline in the Western Mediterranean were the result of internal weaknesses and divisions.

By about 1200, the economies of Western Europe had absorbed most of what Islam and the Orient had to offer. Their astounding feat was their success in pulling ahead in the three centuries that followed. A string of brilliant inventions between 1200 and 1500, despite a temporary setback in the fourteenth century, prepared the way for Europe's eventual technological leadership in many areas. The unique character of European technological change was not only in the ingenuity displayed in making production more efficient, but in the speed with which some of these innovations spread throughout Western Europe.

The windmill combined the ideas of the watermill and the sail. It, too, may have been imported to Europe by Moslems (from central Asia) but in spite of its apparent advantages in arid climates, it was not used widely in the Moslem world.[21] The first windmills that can be documented with certainty were in Yorkshire in 1185. Whatever their exact place of origin, within a few years windmills were used throughout Europe and in 1195 the Pope imposed a tithe on them. Because of the variability of the wind direction in Europe, engineers had to mount their windmills on a pivot so that they could be turned to face the wind at an optimal angle. Later, the introduction of brakes and the inclination of sail beams were introduced to increase stability and efficiency. The windmill owed much to the watermill, adopting its horizontal axle, gearing, and transmission machinery (Reynolds (1983), p. 48). Yet as Usher (1954, p. 176) points out, the windmill was a more complex mechanism than the watermill and its development therefore was slower. The early windmills were post-mills, in which the whole engine house turns towards an optimal position into the wind using a single fulcrum.

Windpower was also utilized more efficiently in shipping as a result of substantial changes in the design of ships. The adoption of the lateen sail in Northern Europe was complemented by the addition of a foremast and a mizzenmast to produce an entirely new type of ship known as a "carrack". Full three-masted rigging combined the advantages of the lateen and the square sails: manoeuvrability in sailing against the wind and speed in sailing with it. The fully rigged ship has been termed "the Great Invention" not

[21] The design of the European mill differed radically from the Asian mills, as it used gears and a horizontal axle. It is thus possible that it was an independent invention. The windmill was reportedly introduced by "German soldiers" (crusaders) into Syria in 1192. Until the twelfth century, the windmill was confined to Persia and Afghanistan (Forbes (1956c), p. 617).

because of its revolutionary nature, but quite simply because "it could do more than any of its predecessors and could do so with considerably less risk" (Unger (1980), p. 219). It was developed first around 1400 in the Basque region around the Bay of Biscay and spread rapidly to Northern Europe and the Mediterranean. At the same time the sails themselves got better: horizontal treadle looms produced higher quality and stronger sails. A second innovation, possibly introduced from China, was the sternpost rudder, which replaced the clumsy steering oars used in antiquity. The rudder appeared in the late twelfth century, and greatly reduced drifting and the physical effort required to keep a ship on course. Ship construction also improved. By about 1300 the so-called "carvel" construction technique was adopted by Northern European shipbuilders. This technique placed boards edge to edge along a skeleton of beams, with caulking between the planks to preserve water-tightness. Carvel-built ships like carracks did not entirely replace ships built by the age-old "clinker-planking" technology, in which the construction was based on overlapping planking. Both methods have advantages: clinker-planking produced much sturdier and heavier ships, because lateral strains were absorbed by the entire body of the ship, whereas in carvel-built ships lateral strain was absorbed by the heavy beams that held the skeleton together. Carvel construction saved on wood and could make much lighter and larger ships than clinker planking; by 1400, ships of 1000 tons were already in existence. Larger ships meant cheaper and faster transportation and improved seaworthiness (Rosenberg and Birdzell (1986), pp. 80–96). Clinker construction survived, however, for smaller ships, and in some instances for medium-sized ocean-going ships as well. By the middle of the fifteenth century the Portuguese "caravel", had emerged. It is this type of ship that was used by Da Gama, Columbus, and Magellan. It was of a carvel construction, had two or three masts, lateen rigged, a sternpost rudder, measuring a burden of perhaps 100 to 200 tons, with a crew of about twenty (See Parry, (1974), pp. 17–22). It was a flexible ship, designed primarily for coastal shipping but sufficiently seaworthy to enable the Portuguese to lead the effort to discover the non-European world, although the caravel eventually was replaced for long-distance voyages by the fully-rigged ship. Yet it should be emphasized that the Europeans were not alone in having the capability to build ships that could cross the oceans: the fifteenth century witnessed a short-lived attempt by the Chinese to explore distant regions.

The Europeans also developed better navigational tools than ever before. There is no evidence that Greek and Roman ships used *any* navigational aides at all, except watching the stars and staying within sight of the coastline wherever possible. Cases like the wreck of St. Paul's ship en route to Rome which completely lost its way because "neither sun nor stars appeared for many a day" (Acts 27:20), must have been common. The Viking sailors, too, used landmarks for direction and often fell victim to *hafvilla,* the loss of all direction at sea. All this changed in the closing centuries of the Middle Ages. The compass, probably an independent European invention that was also known in China, was first mentioned by the English scientist Alexander Neckam in his *De Utensilibus* in about 1180. The device was at first little more than a magnetized needle afloat in a bowl of water. It took until about 1300 for this crude instrument to be turned into a self contained instrument, complete with the sixteen-point wind rose. One specialist (Kreutz (1973), p. 372–73) believes that it was not before 1410 that anyone really steered by a compass. Marine charts and navigating tables of unprecedented accuracy came into use in the thirteenth century, which covered the entire Mediterranean, as well as the Black Sea. In the Mediterranean, the compass was especially important because its deep waters made navigation by sounding (widely practiced in the North and Baltic Seas) impossible. The effects on seafaring were profound. Until the thirteenth century, the Mediterranean was closed for navigation in the winter months; the new navigation technology made year-round shipping possible from around 1280.

In the later Middle Ages, the revival of geography culminated in the translation of Ptolemy's *Geography* into Latin in 1409, which taught Europeans to divide the World into lines of latitude and longitude. To measure latitude, the Portuguese navigators learned to adapt astronomical instruments developed by Hellenistic and later Moslem astronomers. The most famous of these instruments was the astrolabe, an instrument invented by the Hellenistic astronomer Hipparchus for astronomical purposes. The instruments measured the altitude of the polar star, and polar altitude meant that the observer knew his latitude—as had been known since Pytheas of Massalia (third century B.C.). These instruments were often too complex for use at sea, and their first documented use on board occurred in 1456. For navigational purposes, the quadrant, a simplified version of the astrolabe invented by Moslems, was found most useful. Somewhat earlier the cross-staff, which served similar purposes, emerged. In the southern hemisphere, where Europeans ventured after 1450, they had no recourse to the Polar star and used the astrolabe to measure the sun's altitude (Taylor (1957)). Although there was thus progress in Europe, it is far from obvious that by 1450 their navigational techniques were superior to those of China. The difference was that the Europeans were able to keep the momentum of innovation going.

A third area of progress was metallurgical engineering. Mining and metallurgy were especially developed in Southern Germany, Austria, and Bohemia. The landmark achievement here was the construction of larger furnaces that could achieve higher temperatures thanks to water-driven bellows. Higher temperatures permitted the casting of iron, revolutionizing the production of goods requiring the brittle and hard qualitites of cast iron.[22] The blast furnace differed from the primitive bloomeries of antiquity and the Middle Ages in that it generated higher temperatures, which allowed the iron to absorb carbon (about three percent). A higher carbon content lowered the melting point, so that the mixture could actually be poured or cast. The output of this process became known as pig iron which could either be cast into molds, or by subsequent decarburization could be made into the more malleable wrought iron. Although genuine blast furnaces do not appear until the end of the fifteenth century, the improved furnaces appeared somewhat earlier. The blast furnaces produced iron at a far lower cost than the venerable but antiquated bloomeries. On the other hand, most of the crafts of metal working changed little: welding, riveting, and forging remained by and large the same. As far as

[22] Here, too, the Europeans were preceded by the Chinese, who used cast iron "almost as soon as they knew about iron at all" (Needham (1969), p. 101). Needham rejects the idea that Europeans invented cast iron independently, but concedes he has no idea how the transmission was brought about. It is possible that he is mistaken: the application of water power to bellows generated much higher furnace temperatures, which made cast iron almost a natural next stage.

non-ferrous metallurgy is concerned, the discovery of a process that separated silver from copper ores by means of lead around 1450 was the most important discovery of this age.

Metallurgy was important to technical developments in other industries. An example of such a spillover effect can be found in that most famous of all medieval inventions, the printing press. The printing press is justly famous, because it is the first medieval invention for which the name of the inventor and the time of invention are reasonably well established, and for which we know that the inventor singlehandedly solved the entire problem such that within a short period his new gadget surfaced all over Europe. Printing itself was not unknown in Europe by the early fifteenth century. By 1400 playing cards were stamped in Europe, and of course coins had been stamped for two millennia by this time. Yet Johann Gutenberg's invention of moveable type (1453) was an achievement of profound brilliance, made possible inter alia by his knowledge of metallurgy (his father had been goldsmith to the archbishop of Mainz). Casting the movable type was a difficult problem; all letter units had to be of equal length and thickness but varying width. The mechanical solution he found, a mold consisting of two overlapping L-shaped parts, was ingenious. Aided by his metallurgical expertise, Gutenberg used an alloy of tin, zinc and lead for the type, while his molds were made of iron and copper (Cardwell (1972), pp. 20–25). Like the windmill, the printing press spread with dazzling speed. By 1480, there were over 380 working presses in Europe, and in the fifty years subsequent to the invention, more books were produced than in the preceding thousand years.

Of equal technical brilliance was the invention of the weight-driven mechanical clock. Its inventor is unknown, but its occurrence can be dated to the end of the thirteenth century. Earlier in that century, Europeans and Arabs were able to build sophisticated water clocks. But water clocks were unreliable because of freezing and evaporation, and European engineers searched doggedly for a device that would allow them to use the force of gravity.[23] Around

[23] In 1271 Robert the Englishman stated that "clockmakers are trying to make a wheel or disc which will move exactly as the equinoctial circle does, but they cannot quite manage it" (cited by White (1958), p. 122).

the year 1300 the verge-and-foliot escapement mechanism appeared, which succeeded in converting the continuous but variable force exerted by a falling object into the regular oscillating motion required for the accurate operation of a clock. David Landes, in his magisterial work on the history of clocks, refers to it as the Great Invention. Progress was breathtaking: by the middle of the fourteenth century clockmakers such as the Dondi family and Richard of Wallingford were making astoundingly complex devices that indicated not just the time, but also every astronomical motion then known. Moreover, clocks spread rapidly throughout Europe. Landes (1983, p. 57) points to a "clear sense of excitement and pride" in the new mechanisms. Every town felt that it had to possess this marvel to show off to visitors. In White's words, "no European community felt able to hold up its head unless in its midst planets wheeled, angels trumpeted, cocks crew, and apostles, kings, and prophets marched and countermarched at the booming of the hours" (1962, p. 124). In the middle of the fifteenth century spring-driven clocks and watches appeared, as new devices to regulate the uneven force of an uncoiling spring were discovered. The best solution was found to be the fusee, which used a conical axle to equalize the uneven force. The first fusee-driven watches appeared around 1430, and watches became a popular consumer good among the better-off. The advances in clockmaking made the miniaturization of clocks feasible, and thus led to the democratization of time measurement.

Many have stressed the importance of mechanical clocks as both a symptom and a further cause of technological progressiveness. "The clock, not the steam machine," writes Lewis Mumford with some exaggeration (1962, p. 14), "is the key machine of the modern industrial age". It is mechanical, automatic, and demands a high precision level in design and maintenance and thus served as an example to all other machinery. It created order and organization and a shared set of objective information. By the middle of the fourteenth century, the custom of dividing the hour into sixty minutes of sixty seconds each had become standard. Four o'clock was four o'clock for all individuals, an hour was an hour. This communicability of facts and concepts, the "I see what you see" stage of information diffusion, was an important element in the diffusion of innovations. It also permitted, for many purposes, the measurement of productivity. After all, implicit in our notions of efficiency is the need to measure time: productivity is a flow concept. Clocks brought home differences in efficiency: more productive workers and better implements and tools could be seen to produce more output per hour. Productivity comparisons became easier, and with them the choice between the faster and the slower. Clockmakers brought new standards of accuracy and complexity in the construction of mechanical contrivances, and many played important roles in subsequent inventions in other industries.

Another mechanical idea, comparable to the escapement mechanism (though simpler) was the flywheel, which was previously used only for grindstones. Its most important application was the spinning wheel, which first appeared in Speyer, Germany, around 1280. The spinning wheel is also the first known application of a belt-driven transmission. At first the spinning wheels were turned with one hand and the yarn was twisted with the other. By the sixteenth century the foot-operated treadle was added. The spinning wheel greatly increased productivity in the Flemish cloth industry.[24] In weaving, the first mention of the horizontal loom is in the Talmudic comments of Rashi (a Jewish sage living in Troyes in the eleventh century), indicating that this loom was used by professional weavers. It employed a foot-operated treadle that raised and lowered the heddles that opened and closed the "shed" through which the shuttle passed (White (1978), p. 274n). In the thirteenth century, Europeans also mastered the processes of silk throwing, with Lucca in Tuscany becoming the center of the industry. The Lucca silk works made heavy use of water power and complex machinery and may be regarded as the first mechanized textile mills. Some other useful inventions date from the thirteenth century. Spectacles were invented around 1285 in Italy and spread rapidly.[25] Another was the functional button, which brought about a small revolution in apparel design. The Orient for some reason never developed anything like it. The Japanese were delighted with it when they first saw it on the garments of Portuguese traders, and kept the Portuguese name for it (White (1968), pp. 129–30).

---

[24] Adam Smith believed that the spinning wheel doubled the productivity of labor. See Smith (1976), p. 273.

[25] It is telling that Moslem scientists (e.g. Alhazen, who lived around 1000 A.D.) studied the reflection of light in curved mirrors and glass spheres, yet this simple application of optics to better the human lot came from a Western society.

The later Middle Ages also witnessed the increasing use of chemicals in Europe's economy. The Moslems had been better chemists; it was a long time before Europe produced a chemist of the stature of Rhazes or Avicenna. But whatever the Europeans knew and learned about chemicals, they used in production. Alcohol, dyes, alum, saltpeter, mercury, and acids were all used wherever possible and necessary. Gunpowder may not have been a European invention, but the Europeans soon designed and built guns that left Islam and the Orient at their mercy unless they were willing to follow suit. But most chemicals were used for peaceful ends, such as staining windows, dyeing, tanning, oil paintings, medicine, and metallurgy. It was progress without science, chemicals without chemistry, but it worked. Progress was attained by thousands of forgotten tinkerers and craftsmen, often replicating each other, many of them wasting their creative energy in the fruitless pursuit of alchemy and other dead ends. Yet progress there was: slow perhaps, but inexorable in the long run.

By 1500, Europe had achieved, approximately, technological parity with the most advanced parts of the Moslem and Oriental worlds. Indeed, in the assessment of some historians, by that time Europeans already controlled more energy, machinery, and organizational skill than any civilization, ancient or contemporary (Lach (1977), p. 400). Europe was soon to turn from borrower to lender. Much of this achievement preceded the beginning of European science. Systematic learning had little to do with technological progress. Medieval technology differed from classical and modern technology in another important respect. Cardwell (1972, pp. 9–10) has pointed out that unlike classical technology, it was not grandiose and extravagant. Apart from a few imposing church buildings and castles, it was mostly concentrated in the private sector. It was carried by peasants, wheelwrights, masons, silversmiths, miners, and monks. Western technology was, above all, practical, aimed at modest goals which eventually transformed man's daily existence. It produced more and better food, transportation, clothes, gadgets, and shelter. It is of this stuff that Schumpeterian growth is made.

4. THE RENAISSANCE AND BEYOND: TECHNOLOGY 1500–1750

As noted, by 1500 Europe was no longer the technological backwater it had been in 900 or the upstart imitator of 1200. It is clear that Europe owed China a great deal, as Joseph Needham has tirelessly argued.[26] Yet in the two centuries before 1500, Europe's technological creativity had become increasingly original. In the later Middle Ages Chinese technology had become in Landes's phrase, a "magnificent dead end." After 1500 China ceases to be of much interest to the historian of technology. Its use of iron and water power did not lead to a Chinese Manchester any more than their knowledge of printing led to a massive outpouring or printed books in China or than Su Sung's famous water clock led to a large clock to be erected in the center of every town in China.[27]

In the centuries after 1500, the gap between Europe and the rest of the world gradually widened, even though it was an age marked by relatively few major inventions in Europe. Technical progress in the conventional sense continued unabated. The increase in productivity, however, became more gradual and consisted largely of clusters of minor improvements and modifications to existing techniques. One explanation of the absence of discontinuous breakthroughs between 1500 and 1750 is that although there was no scarcity of bold and novel technical ideas, the constraints of workmanship and materials to turn them into reality became binding. If one insists on dating inventions according to the first time they occurred in the human mind or on paper, this period may indeed be regarded as the equal of the Industrial Revolution. But the paddle-wheel boats, calculating machines, submarines, parachutes, fountain pens, steam-operated vehicles, power looms, and ball-bearings envisaged in this age—interesting as they are to the historian of ideas—had no economic impact because they could not be made practical.

From a purely economic point of view, the most important technological change in terms of its potential contribution to material welfare can hardly be termed an invention at all. The "new

[26] Much of Needham's tireless defense of the Chinese origins of Western technology is a needed antidote to the Eurocentric histories of technology that fortunately are now falling out of style. But Needham may have exaggerated the Chinese influence. It may well be true, as a widely quoted phrase notes, that "the world owes far more to the relatively silent craftsmen of ancient and medieval China than to the Alexandrian mechanics" (Needham (1969), p. 58). But by 1500 Europe already owed far more to its own silent craftsmen and engineers.

[27] Europe's debts to the Orient were not confined to China. India, Tibet, and even Malaya contributed to European power technology, metallurgy, and textiles (White (1978), pp. 43–58).

husbandry" as it is now called was a set of modifications in agricultural practice that made its first appearance in the Low Countries by the closing of the Middle Ages. It spread, ever so slowly, to England and eastward, but by 1750 their adoption was far from complete and in some areas, including most of France, had hardly begun. Yet the principles of the new husbandry were revolutionary, and its adoption led eventually to huge increases in agricultural output. The three elements of the new husbandry were all closely related: stall-feeding of cattle, new crops, and the elimination of fallowing. The result was that farmers were able to maintain more and better-fed cattle, thus increasing the supply of animal products. Better fed animals produced more fertilizer, which helped to increase cereal yields. The new fodder crops such as alfalfa, clover, artificial grasses, turnips and mangel-wurzels also turned out to be useful as alternating crops to cereals in new rotations. Some of these crops were nitrogen fixers and all of them broke disease and pest cycles. Thus, with the increased supplies of fertilizer, it was found that the fallowing of land became less necessary and gradually started to disappear in a few regions. The new husbandry was a tale of complementarities, of mutually reinforcing and symbiotic changes, but it was slow to unfold. Some of the new crops were not suitable to heavy clay soils, others needed better drainage than was available. Capital scarcities, the scattering of plots, and hostility to novel practices by those who were frightened or threatened by them, slowed down diffusion. Some scholars believe that the adoption of the new husbandry depended on the enclosure of open fields but this is now disputed. In any event, the often-used term "agricultural revolution" to denote the introduction of the new husbandry is misleading: there was nothing abrupt about it. Yet in the long term its occurrence had profound consequences for the economic well being of most Europeans.

Moreover, although it permitted larger quantities of food to be produced, technological progress in farming did little to reduce the toil of the men and women working in the fields. Even the new implements introduced at this time were, by and large, capital- and land saving rather than labor saving. The seed drill is a case in point: in traditional European agriculture, sowing was carried out by "broadcasting," spreading the seeds by hand. This technique not only wasted seed and led to an uneven utilization of the soil, it also made weeding difficult because of the uneven germination of the plants. In the sixteenth century the practice of "setting" seeds (by using sticks to make holes in which seeds were dropped) was known. The modern seed drill, which greatly increased the yield–seed ratio, was associated with Jethro Tull, who built and demonstrated the first prototypes around 1700, though the implement was little used before the nineteenth century. Tull also suggested the introduction of horse hoeing in 1714. New iron plows, introduced first in the Low Countries and then in England, also probably saved capital more than labor. The new plows reduced the friction with the soil by curving the shape of the mouldboard that cut the furrows. It was difficult, however, to shape wood into the exactly desired shape, and so after 1650 mouldboards were increasingly made of iron. This led to the disappearance of wheels and the reduction of the number of draft animals required for plowing. In 1730 the Rotherham or Dutch plow was patented in England.

In the area of energy use, medieval techniques were improved but not revolutionized. The windmill continued its winding road to ever greater efficiency when Dutch and Italian engineers in the sixteenth century introduced the tower mill, which left the structure permanent and made the axis and roof pivot toward an optimal angle. By the seventeenth century the windmill supplied the Dutch economy, at that time the economic *wunderkind* of Europe, with a cheap, clean, and inexhaustible source of energy that is the envy of today's ecologists. The Dutch were often able to increase the efficiency of manufacturing using windpower through technical ingenuity, as was the case with the *Hollander,* a device used in papermaking invented around 1670. It consisted of horizontal rollers with spikes and mallets which literally ripped up the rags used for pulping (Hunter (1930), pp. 170–71). Consequently, Dutch paper was of a higher quality than that made elsewhere in Europe. Water power generation and transmission, too, became more sophisticated. Yet in spite of the attractive features of wind-and-water power, there was a perceptible need for a machine that would not depend on *current* supplies of wind and water power and thus make it independent of the vicissitudes of European weather.

On or immediately below the surface of Europe lay vast supplies of stored-up solar energy in the form of peat and coal. The use of peat and coal in Europe was not new in 1500, but its geographic expansion and increasing number of uses have prompted some to assign it an importance equal to the greatest inventions of the Industrial Revolution.[28] In Britain and in a few places on the Continent, including the Principality of Liège in what is today Eastern Belgium, coal was used in iron forges, glassmaking, saltmaking, soapboiling, alum production,[29] and lime burning. The fuel-intensive brewing industry could not use regular coal, as it ruined the taste of the beer, but British brewers learned to use charred coal later known as coke. Similarly, the Dutch used their abundant peat supplies extensively for home heating as well as for a myriad of industries that helped create the Dutch golden age. Brickmakers, madder producers, kiln operators, salt refiners, bakeries, bleachers, tile makers and many others made extensive use of peat. Only two major uses of fossil fuels eluded this age: the smelting of iron and the efficient conversion of thermal energy (heat) into kinetic energy (work).

The period 1500 to 1750 saw major improvements in the use of blast furnaces. Their size and efficiency increased substantially: in 1500 a blast furnace could produce 1200 kg a day, by 1700 this figure had increased to over 2000 kg, reducing fuel consumption in the process. An important improvement was the adoption of a continuous smelting process, in which ore and fuel were fed into the furnace continuously, producing a continuous flow of "pig" iron. Such a "found-day" could last up to forty weeks by 1700. This period also saw the use of reverberatory furnaces, first described by the Italian Vanoccio Biringuccio in his *De la Pirotechnia* (1540) and applied to the English glass industry around 1610. These furnaces eliminated the chimney and used underground pipes to draw in fresh air. A dome-shaped roof lined with refractory clay reflected the heat back into the oven, generating very high temperatures. Another improvement was the *trompe*, dating from the mid-sixteenth century, which used flowing water to blow air into the

[28] Consider John U. Nef's statement in a classic article: "By the mid-seventeenth century a new industrial structure was being built in England on coal, and this structure provided the basis for the industrialized Great Britain of the nineteenth century" (1964, p. 170).

[29] Alum is a double sulfate of ammonium and a trivalent metal, used widely in a range of chemical industries including dyeing, tanning, papermaking, and pharmaceuticals.

forge like a reversed air lift pump. Yet there is some evidence (Smith and Forbes, (1957), p. 30) that these technical improvements were insufficient to prevent rising fuel and labor costs from raising the price of iron.

Economic activity expanded not only sideways but also downward into the earth itself. From about 1450, European mining entered an age of progress unlike anything ever seen before. Here, too, we have no famous inventors, just an endless succession of anonymous inprovements on the margin. We do have, however, a hero of sorts, namely Georg Bauer, who, under his latinized name of Agricola, wrote *De Re Metallica,* published posthumously in 1556, which is one of the finest and most detailed books on mining engineering ever written. From it we can infer the improvements introduced into mining after 1450. Agricola describes the machines used for drainage and ventilation, the cranes, the construction of shafts, even the sampling of ore quality. The technical problems in mining appear to be universal: flooding and explosions lead the list. Germans led Europe and the world in mining technology, developing the transmission of water power to high elevation mines from waterwheels in the valleys by means of overland rod systems, the use of rails for underground transport, horse-operated treadmills to run windlasses, and above all a variety of pumping devices (which were subsequently applied to fire fighting and other uses.[30] Of comparable influence was the Bohemian mining engineer, Lazarus Ercker, whose magnum opus on mineral ores and mining techniques published in 1574 was used for generations as a manual on assaying and sampling. Agricola and Ercker were both empiricists, not scientists. There was no theory in their work, just descriptions of things that worked.

The large number of technical "how-to" books published after 1450 provided a vehicle through which technology was diffused through Europe. Renaissance engineers wrote about a variety of machines and contraptions, many of them serving architectural and

[30] It is with only little exaggeration that one historian refers to this period as "the age of the pump" (Burstall, (1965), p. 144). Some of the writers on technical matters of the time devoted much attention to describing a variety of pumps. In addition to Agricola, there was Agostino Ramelli, whose enormous book *Dell' Artificiose Machine,* published in 1588, contained descriptions of over a hundred pumps.

military purposes. A technical literature emerged, written by engineers for engineers, and technical knowledge became increasingly communicable and thus cumulative. One of the earliest and greatest of these figures was the Sienese engineer Marianus Jacobus Taccola, whose *De machinis libri* summarized the state of the art in machine technology in the middle of the fifteenth century. His influence was such that he was often called "the Sienese Archimedes." This literature also illustrates the growing respect shown by Europeans for machines and the people who made them. Yet outside a few areas, it is unclear that these writings had much effect on the industrial practices of the time (Gille (1966)). All the same, during the centuries of the renaissance, there was a demand for technological information like never before. Jacques Besson's *Theatrum Instrumentarum et Machinarum,* published in Latin and French in 1569, went through three translations and seven editions in the following thirty five years.

Among the successes of Renaissance technology were its achievements in hydraulic engineering, an area in which the Classical World had excelled, but which had been neglected for many centuries. Italian engineers such as Leonardo Da Vinci wrote extensively about dams, pumps, conduits, and tunnels. The leading practical hydraulic engineers of the time were Dutch. After the disastrous floods of 1421, Dutch engineers gradually reclaimed their land, protecting it from the large rivers and the sea. Dutch engineers used the experience they accumulated to help solve similar problems in the marshes of Poitou and to drain the English Fens in the seventeenth century. From Venice to Lübeck, hydraulic engineers struggled with the oceans, rivers, and swamps. One of the greatest engineering feats of the Baroque era was the Marly pumping plant, built between 1678 and 1685 to provide water to the royal palaces at Versailles, Marly, and Trianon, by the Walloon engineer Arnold de Ville. Fourteen huge water wheels drove 221 pumps, and while the project had flaws (the pumps were inefficient and noisy), it still stands out as a triumph of seventeenth century hydraulic engineering.

In textiles, the spinning wheel was provided with a treadle and a crank that increased its ease of operation. The foot-operated drive released both hands of the operator to tend the yarn. The stocking frame, invented by William Lee about 1589 and diffused through

Europe in the first half of the seventeenth century, counts as one of the few true "major inventions" of this age. The Dutch loom, which could weave simultaneously up to twenty four ribbons, dates from 1604, its invention attributed to a Dutchman by the name of William Dircxz Sonnevelt. An important innovation of the time was the manufacture of new fabrics. It is in this period that Europeans first started to make cotton products themselves (rather than importing them from the Orient). With the predominance of cotton still far in the future, the early modern period saw a great expansion of the production of worsteds, a woolen product made of coarse wool that had been combed rather than carded. It did not need fulling, which made it relatively cheap, and was lighter than wool, making it more attractive to consumers in warmer climates. Although no great technical breakthroughs were made, the techniques spread from the Continent to Britain, where it became a major part of the industrial sector known as the "New Draperies."

The geographic discoveries were in many ways the dominant feature of this age. In some ways, the discoveries may have slowed down the rate of technical progress as they absorbed much of the energies of the more adventurous and resourceful Europeans. Yet technological and geographical discovery were often complementary, and the interplay between Smithian Growth and Schumpeterian Growth took many forms. One of these was the design of ships. Although this age did not see anything like the great changes in shipbuilding and seafaring techniques of the later Middle Ages, it managed to succeed in less spectacular but economically crucial aspects, namely to reduce the cost of transport significantly. The Dutch shipbuilding yards were at the center of progress. In 1570, a Dutch sailor came up with the idea of a separate topmast, fitted in a cap on top of the mainmast. The advantage was twofold: the tall and expensive spartrees used to make mainmasts could be dispensed with, and the topmast could be removed in bad weather reducing the chances of damage to the mainmast (Unger (1978), p. 28). The Dutch shipbuilding industry led the movement toward specialization, with different types of ships made for different functions, with at least thirty nine different types of canal- and river boats alone. The culmination of this trend was the Dutch *fluytschip* or "flute," according to tradition first built in 1595, but in fact the crowning achievement of a century of continuous rationalization and improvement. The *fluyt* was cheap to build and operate, carried a small crew and large cargoes, and thus enabled the Dutch to undercut the French and British carrying trade by 30–50 percent, until Holland's competitors started to imitate her (Derry and Williams (1960), pp. 209–210). With the use of heavy naval guns, ships in Northern Europe became specialized again, as they had been in Roman times and in the medieval Mediterranean. Cargo ships and naval vessels were differentiated, and large unarmed merchant ships sailed in convoys, protected by heavily armed men-of-war (Unger (1981)). In navigation, too, improvement was marginal, with better quadrants and maps, but the greatest difficulty seamen faced—determining accurate longitude while at sea—was not satisfactorily solved until the eighteenth century.[31]

Where the discoveries had a clear and visible impact was in the introduction of crops from other Continents into Europe or, failing that, their cultivation elsewhere by European entrepreneurs for the sake of European consumers. Three crops originating from the New World were brought to Europe: maize (corn), tobacco, and potatoes. Tobacco was grown successfully in many places in Europe, though the quality rarely matched that of the best varieties grown in America. Maize was grown in southern France and Italy as a lower class food. The potato eventually had the greatest impact on the European economy as a nutritious and cheap food, first in Ireland, then in the Low Countries, and after 1800 throughout most of Europe. Similarly, the Europeans were exposed to new industrial goods which led to the eventual growth of import substitutes. One of these was the Chinaware industry which started in Saxony in 1712; another was the British cotton industry which arose from a desire to compete with the high quality Indian cotton goods known as calicoes.

The age of discoveries was thus the age of exposure effects, in which technological change primarily took the form of observing

[31] The problem was essentially one of building an accurate clock that would withstand the motion on sailing ships, so that one could compare the time on board (measured from solar charts) with the time at a given point like Greenwich; a comparison between the two allows the inference of longitude. In 1714 a specially established Board of Longitude promised the huge sum of £20,000 to "such person as shall discern the longitude at sea." A satisfactory clock was built by John Harrison in 1762.

alien technologies and crops and transplanting them elsewhere. The aggressive Europeans not only adopted crops from America in exchange for the livestock, wheat, and grapes they transplanted into the New World. They also performed a scarcely less aggressive function in transplanting non-European flora from America to Africa and Asia and back in a massive act of what could be called ecological arbitrage.[32] Thus they introduced bananas, sugar, and cotton in the new world, and cassava (also known as manioc) in Africa where in the nineteenth century it became the staple crop in many areas, although not without much resistance (Crosby (1972), p. 187). Sweet potatoes and peanuts were a great success in China after Portuguese traders brought them to Asia in the sixteenth century.

The discoveries, together with improved technology, also increased the supply of fish, an important part of the European diet. Fish is rich in protein, and exactly because domesticated animals were expensive to raise and the supply of game small in most parts of Europe, fish was indispensable. The problem was one of preservation: in an age of slow transport and no refrigeration, fresh fish could only reach those living close to sources of supply. Herring was the main catch in Europe, first in the Baltic and later in the North Sea. In the late fourteenth century, Dutch fishermen discovered the technique of gutting and salting fresh herring, which allowed preservation for long periods. By about 1415, they had introduced drift nets that were towed along ships and increased catches substantially. The ships that carried these nets, introduced at about the same time, resembled little floating factories. They carried coopers and salters aboard who processed the fish immediately. These fishing ships were known as *busses* and helped the Dutch establish a domination in North Sea fishing that lasted for centuries. The other major crop provided by the sea was cod, which was gutted and dried to produce the famous "stockfish" sold throughout Europe. The discovery of huge supplies of codfish off the banks of Newfoundland in 1497 by John Cabot, and the use of a new type of line with thousands of hooks, gave the Europeans a new and

[32] The ecological consequences of the age of discoveries have been explored masterfully by Alfred Crosby (1972, 1986).

unexpected free lunch of dried cod, not appetizing perhaps by modern standards, but rich in proteins.

Above all, the age of discoveries was one of instruments. Instruments in Western technology came before machines. The affinity of Europeans for gadgets derived to a large extent from the clockmaking industry. Clockmakers taught others the wonders that precision-built spring-driven gears and cogs could achieve. By the middle of the fifteenth century the German town of Nuremberg had become the world's center for gadgets. Its fame is immortalized by E. T. A. Hoffmann and Jacques Offenbach in the tale of the Doll of Nuremberg. Not all instruments were toys. Astronomical instruments and compasses were crucial to the world-wide navigation in which Europeans became involved.[33] Military technology required precision for the calibration and sighting of guns. Commerce required precision scales, real estate odometers. A special branch of the instrument making industry was optics. The earliest spectacle lenses were convex and could only aid the long-sighted. A little before 1500, concave lenses were developed that corrected short-sightedness as well. The telescope was invented by Dutch opticians in the early 1600s. Though of limited direct economic use, the telescope nicely illustrates the pragmatic bent of the European's mentality at that time. Within a few years, Prince Maurice of Nassau used the telescope to gaze at the Spanish armies and his sea captains used it to look for cliffs and hostile galleons at sea, while in Padua a mathematician by the name of Galileo used it to gaze at the moons of Jupiter. Technical ideas and gadgets that worked, and worked well regardless of the environment, spread more rapidly than ever before.

The precision instrument industry produced important spillover effects to the manufacturing sector. The main breakthroughs had to await the Industrial Revolution, but the lathe, one of the oldest

[33] From Nuremberg and Augsburg the art of instrument making spread to Louvain and from there to London. The London instrument maker Humfray Cole was apprenticed to the Liege craftsman Thomas Gemini. Among Cole's customers were Francis Drake and Martin Frobisher. Gemini himself had studied in the South of Germany. Another German, Nicholas Kratzer, lived in England for many years. It is interesting to note that although the Germans themselves were not much involved in discoveries, the instruments they made were. The European world by 1500 had become sufficiently integrated that knowledge and technique could spread across boundaries without difficulty.

carpenter's tools in use, underwent improvements as clock- and instrument makers needed precision parts and accurately cut screws, and opticians needed precision-ground lenses. In the sixteenth century the flywheel and the crank, those irrepressible medieval ideas eternally in search of applications, were applied to the lathe. The greatest lathemaker of the age was Jacques Besson, a French engineer employed by the court of King Charles IX, who built, around 1569, an ingenious and sophisticated screw-cutting machine. The Besson machine was semi-automatic in that the operator needed only to pull and release a cord. Consequently it was rather complex and probably not widely used (Woodbury (1972), p. 57).

Instrument making in the sixteenth and seventeenth centuries was an art, not a standardized technique. Most improvements were the result of serendipity and trial and error searches. Learning and training took place mostly through apprenticing and informal contact. Mechanics had to build their own parts, and often the gap between the visionary who saw what *might* be done and the craftsmen whose material and tools limited what *could* be done was too wide to be bridged. The most famous of these visionaries was of course Leonardo da Vinci, whose mechanical brilliance was on a par with his other talents. Leonardo left us with five thousand pages of unpublished notebooks, much of which dealt with machinery. Yet the Last Supper notwithstanding, Da Vinci's technical creativity produced few free lunches, and his technical insights were rarely realized in his lifetime. Neither, for that matter, did the equally prophetic technical dreams of Leonardo's precursor, Francesco di Giorgio (1439–1502).[34] As we shall see, the Industrial Revolution became possible when mechanics and machine tools could translate the ideas and blueprints into accurate and reliable prototypes. Until then instruments and tools were handmade, expensive to make and repair, and limited in their uses.

The age 1500–1750 is better known for its scientific achievements than its technological breakthroughs. The interaction between the two is the subject of an extensive literature and cannot be done any

[34] Not all of Leonardo's brilliant ideas were of academic interest only. His invention of mitred gates, fitted with small wickets for letting the water through, was applied to the Milan canal, and then to the Exeter canal in 1564.

justice here. What is striking, however, is that during the centuries of the Renaissance, the classical dichotomy between thinkers and makers all but disappeared in Europe, whereas the modern distinction between scientist and engineer had not yet appeared. Many scientists made their own instruments and contributed to the solution of practical problems associated with their manufacture. Galileo built his own telescopes and supplemented his salary as a professor at the University of Padua by making and repairing instruments. Robert Hooke, the brilliant and eccentric physicist and biologist, pioneered the use of balance springs for watches and invented the so-called Hooke's joint, an elegant device for power transmission. Together with Robert Boyle, one of the most versatile scientists of his age, he produced a superior airpump, originally invented by another scientist of fame, Otto von Guericke. The physicist and mathematician Christiaan Huygens invented the pendulum clock and was one of the first to suggest an engine using an internal combustion chamber for which he wanted to use gunpowder. Leibnitz worked on a wind-driven pump to remove water from mines in the Harz mountains. Even Isaac Newton, despite his professed (and unusual) lack of interest in technology, made a contribution to the perfection of the marine sextant (Gille (1969), p. 314).[35] Thus, scientists were more important to technological change than science itself.

The age of the Renaissance and Baroque also witnessed the beginning of the application of mathematics to engineering in a variety of areas. During the Middle Ages, Europe had made few major contributions to mathematics. The Arabs adopted and perfected the decimal numeral system that still (unjustly) bears their name. They learned algebra from India and preserved and extended classical geometry. In the later Middle Ages the Europeans first saw, then learned, then imitated, then applied, then improved, and eventually took the field over so that modern mathematics is by and large a European product. Mathematics was discovered to be useful to all economic activity, not just engineering. The use of the Arabic

[35] One technological historian refers to seventeenth century science as "experimental philosophy," an oxymoron that nicely captures the essence of advances in knowledge in those years. As late as the closing years of the eighteenth century, the physicist Count Rumford asserted, with some exageration, that "invention seems to be particularly the province of men of science" (cited in Hall (1967), p. 115).

numerals was first introduced to Europeans by Leonardo Fibonacci of Pisa, whose *Liber Abaci* was published in 1202. The system diffused slowly, but it had huge advantages in accounting, measuring, and calculating and it was doubtlessly instrumental in the development of double-entry bookkeeping, which appeared in the middle of the fourteenth century.[36] By the sixteenth century, Europeans started to discover new ways to make mathematics more useful in daily life. Simon Stevin, a Flemish engineer, suggested in 1585 the use of what was to become known as the decimal point. About a decade later the Scottish mathematician John Napier discovered logarithms, making business calculations like compound interest easy and accurate. New areas of application turned up repeatedly. William Oughtred, the Rector of Albury in England, invented the slide rule in 1621. John Graunt, an English merchant, published the first life tables in 1662 and should be considered the founder of demography. Soon afterwards the political arithmeticians William Petty and Gregory King pioneered the idea of observing a society through aggregate statistics.

The main use of mathematics was in mechanical engineering. Mathematicians and engineers discovered that they needed each other. Mathematics was needed in measurement, civil engineering, ballistics, optics, navigation, and hydraulic systems.[37] Clockmakers wanted to know the optimal shape of the teeth on gear wheels that would minimize friction: the mathematicians Ole Roemer and Christiaan Huygens were able to show that the epicycloid was the curve satisfying this condition. In other areas such as shipbuilding and machinery, the application of mathematics was more difficult, because much of the mathematics still needed to be developed itself. Galileo's development of mechanical physics and the later invention of calculus were necessary for these further advances.

As Cardwell points out, Galileo's theory of mechanics is interesting to the economist because the concept governing it is one of

[36] The papers of the merchant of Prato, who left his estate to his home town in 1410, include an astonishing amount of accounting material, including five hundred account books, which bear witness to the usefulness of the decimal system (Burstall (1965), p. 112).

[37] "Mechanics is the paradise of the mathematical sciences, because it is in mechanics that the latter find their realization . . . when none of the mathematical sciences can be applied, there is no certitude," wrote Da Vinci (cited by B. Gille (1969), pp. 135–136.

efficiency: "The function of a machine is to deploy and use the powers that nature makes available in the best possible way for man's purposes . . . the criterion is the amount of work done—however that is evaluated—and not a subjective assessment of the effort put into accomplishing it" (Cardwell (1972), pp. 38–39).[38] In the writings of Galileo, the leading scientist of his time, economic efficiency is linked with science. In this, he differed radically from his inspiration, Archimedes, and this difference between the two scientific giants who established the science of mechanics epitomizes the difference between classical and early modern society. It is hard to imagine how the great engineers of the Industrial Revolution could have advanced machine making beyond the purely empirical without a unified theory of machines.

The period between 1500 and 1750 was thus one of technological development, but not one of revolutions. Considering the handicaps and obstacles that new technology faced in this age, it is surprising that the process did not grind to a halt altogether. As noted, the great discoveries may have been a substitute for technological frontiers: the challenging and possibly enriching opportunities lay overseas. Moreover, the effects of the Reformation on the rate of technological progress were probably on the whole negative. That Protestantism itself was conducive to technical changes is doubtful. What matters to innovation is not only what one believes per se but to what extent society tolerates deviation and non-conformism (Goldstone (1987)). The Reformation, and its natural sequel the Counter-reformation, made Europe a more bigoted place than it had been since the crusades: Giodano Bruno was burned by the Catholic inquisition, Miguel Servetus by its Calvinist counterpart in Geneva. Throughout Europe in the sixteenth and early seventeenth centuries, the patience of the authorities for people who thought for themselves and were critical of dogma was wearing thin. This pressure to conform slowed down technical change, though it is difficult to asses to what extent. In southern Europe, which came increasingly under the domination of the reactionary power of the Counter-reformation, the climate for technological creativity

[38] Galileo wrote in his *Motion and Mechanics* that the advantage of machines was to harness cheap sources of energy because "the fall of a river costs little or nothing."

changed for the worse.[39] Moreover, religious differences helped trigger wars that destroyed some of the most active centers of technological change in Europe, especially in the Southern Netherlands (1568–1590) and most of Germany (1618–1648). The ability of Europe to maintain its momentum, despite worsening circumstances, serves as a testimony to the resilience of the forces of technological progress. If Antwerp and Augsburg were destroyed, there were always Amsterdam and London.

Between 1500 and 1750 important changes in the form of industrial organization occurred in Europe, and these changes were likely to have affected the rate of technical progress. One was the de-urbanization of industry. Cities had always been unhealthy places, plagued by epidemics, poor water supply and sewage, fires, and overcrowding. During the religious wars, many cities were besieged and sacked. The poor quality of life and short life expentancy in cities raised urban wages relative to the countryside, since towns had to continue to attract migrants. Another cause behind manufacturing's move to the countryside was the tight corporate structure of craft guilds, which restricted entry and imposed strict rules on the quality and price of output. It may well be the case that by the sixteenth century the guild system in the towns had begun to stifle technological progress to protect its monopolistic position. There are repeated cases of the authorities trying to suppress innovations, doubtlessly instigated by lobbies of vested interests.[40]

---

[39] Around 1588, Giambattista della Porta constructed in Naples a successful incubator for hatching eggs. His innovative proclivities drew the wrath of the holy inquisition, who threatened to persecute him as a sorcerer. Porta wisely gave up his experiments, and they were resumed later by the Dutchman Cornelius Drebbel and the Frenchman René Réaumur, in societies where non-conformist ideas were better tolerated.

[40] As early as 1397, Cologne tailors were forbidden to use machines that pressed pinheads. In the fifteenth century, the Paris scribes' guild was able to delay the introduction of printing by decades. The ribbon loom was invented in Danzig in 1579, but its inventor was reportedly drowned secretly by orders of the City Council. Twenty five years later the ribbon loom was reinvented in Holland and thus became known as the Dutch loom, though in Holland too resistance to it was stiff. In France the "great printers revolt" in the sixteenth century was triggered by labor saving innovations in the presses. A century and a half later, a similar fate befell John Kay, the inventor of the flying shuttle. But the prolonged opposition of vested interests against the flying shuttle in Britain was ineffectual.

Entrepreneurs discovered that rural labor was cheaper and more docile, and began to "put out" work to the countryside, where peasant families spent much of the slack season spinning, weaving, frame-knitting, and making nails, needles, baskets, cutlery, shoes, toys and so on. The cottage industries were confined mostly to low-skill tasks. Finishing, design, and the production of high-quality or complex goods remained in the realm of specialists. Cottage industries were not altogether impervious to technical change, however: the spinning wheel, the mainstay of cottage industry, was repeatedly improved, and Lee's stocking frame also found widespread use among domestic manufacturers. In 1733, John Kay, a Lancashire mechanic and weaver, invented the flying shuttle, which greatly increased the productivity of hand loom weavers.

The rise of nation states between 1450 and 1750 also had important effects on technology. Although government officials rarely participated directly, many governments followed deliberate technological policies. The objectives of these policies were, of course, frequently political and military such as the design of fortifications, the casting of cannon, and the construction of men-of-war. Yet, in this period a mercantilist outlook led governments to follow an active industrial policy. States increasingly employed and subsidized engineers, and awarded monopolies, patents, and pensions to inventors deemed to have made important contributions to the welfare of the realm. When a nation felt left behind, it sometimes made a deliberate effort to catch up. The most famous example of government-inspired technological diffusion is Czar Peter the Great's stint as a carpenter in a Dutch shipyard. More than two hundred years earlier, however, the Czars were already sending for architects, miners, printers, and metal workers from the West. Similarly, King Gustav Adolph of Sweden invited the Dutch industrialist, Louis De Geer, a Walloon by birth, to set up blast funaces in Sweden, which—thanks to her quality ores and large forests—soon became a leading iron producer. These initiatives were not confined to the peripheral areas of Europe. The Lyons silk industry was founded when Louis XI enticed some Italian craftsmen to settle there, and the Sforza family attracted some of the best engineers of the time to work in Milan in the first half of the sixteenth century.

Some European governments learned that protecting the property rights of the inventor encouraged technical change. The idea of a patent as a temporary monopoly position to reward inventive activity emerged from customs in mining activity. Mining contractors were awarded monopoly rights in discoveries of new mineral resources. The arrangements were subsequently adopted in other activities such as grain milling and eventually applied to what we could today call "new inventions." This custom appears in Northern Italy in the first quarter of the fifteenth century. In 1474 a formal patent system was enacted in Venice the preamble of which noted that if "provisions were made for the works and devices discovered by men of great genius, so that others who may see them could not build them and take the inventor's honor away, more men would apply their genius . . . and build devices of great utility to our commonwealth" (Kaufer (1989), pp. 1–7). By the middle of the next century the idea had penetrated throughout much of Europe. The most effective and famous patent law was the Statute of Monopolies in England, passed in 1623.[41] States also sponsored scientific societies, such as the Royal Society in Britain (chartered 1662) and the Académie Royale des Sciences in Paris (1666). These societies soon gravitated toward pure science, and the more important societies from the point of view of "facilitating the manual arts" (as Robert Hooke, one of the founders of the Royal Society, put it) were private.

Despite the absence of Great Inventions, then, the centuries of the late Renaissance and Baroque were an age in which western society became permeated with technology. As Bertrand Gille (1969, p. 146) points out, everything about the Renaissance was technological, including its art and its political philosophy. Western people were becoming conscious of the infinite possibilities that technology promised for human welfare, and realized that by accepting change as a way of life, they could have access to a never-ending stream of free lunches.

5. THE YEARS OF MIRACLES: THE INDUSTRIAL REVOLUTION 1750–1830

By 1750, Europe had consolidated its technical superiority over the rest of the world. The Oriental empires, from the Bosporus to

[41] According to Bertrand Gille (1969, p. 146) more patents were granted in Germany in the sixteenth century than in the eighteenth century.

Tokyo Bay, were falling behind by isolating themselves from the West and experiencing a slow-down in their own technological progress. Some of them, like India, were already coming under Western domination. Yet it seems plausible that if European technology had stopped dead in its tracks—as Islam's had done around 1200, China's by 1450, and Japan's by 1600—a global equilibrium would have settled in that would have left the status quo intact, with few elements able to upset it. Instead, the past two centuries have been a period of ever-accelerating change, a disequilibrium of epic proportions that cannot be compared to anything that came before it. In two centuries, daily life changed more than in the seven thousand years before. The destabilizing agent in this dizzying tale was technology, and Western technology alone.

As emphasized above, technological change did not begin with the Industrial Revolution. The difference between the period after 1750 and the period before was one of degree; but degree was everything. The effects of the gains in productivity allowed Europe to expand its population manifold, to extend, for a while, political control over most of mankind, to provide Europeans with a quality of life that was light-years away from "traditional" societies, and to reshape technology elsewhere in the European image.

In recent years, the concept of the Industrial Revolution has come under serious scrutiny. Authors such as Jones (1988) have argued that there was little economic growth in Britain in the second half of the eighteenth century. Since it has been customary to identify the Industrial Revolution with the growth of income per capita (see Wrigley (1987) for a recent restatement), the implication seems to be that the concept of an Industrial Revolution is dispensable. Yet recent calls to ban the concept from our research papers and lecture notes seem misplaced to say the least, because the Industrial Revolution was not primarily a macroeconomic event which led to a sudden acceleration of the rate of growth, although growth eventually became an inevitable corollary of it. The definition of the Industrial Revolution as economic growth suffers from a number of fatal defects. First and foremost, by focussing on a per capita variable it glosses over changes in a ratio in which both numerator and denominator were increasing more or less *pari passu*. As it happened, the years of the Industrial Revolution tended to be years of rapid population growth, and per capita changes were swamped by demographic changes. Second, economic growth need not be a result of industrial change at all, it could be (and often was) rooted initially in agricultural or commercial developments. Third, income per capita is notoriously hard to measure over a period in which the economy undergoes rapid changes in the way the market operates. Commercialization implies that goods previously produced by the household are now purchased on the market. Unless meticulous adjustments are made, these changes tend to bias the measures of economic growth. The appropriate way to think about the Industrial Revolution remains, therefore, as a process of accelerating and unprecedented technological change. In the words of T. S. Ashton's famed schoolboy, it was first and foremost a "wave of gadgets," a string of novel ideas and insights that made it possible to produce more and better goods and do so more efficiently. To return to our earlier definition, a clustering of macroinventions occurred, leading to intensified work in improvement and adjustment and thus creating a complementary flow of microinventions. The outcome was that output per capita increased little initially, but that in many industries and sectors production technologies changed dramatically, preparing the way for sustained Schumpeterian growth in the second half of the nineteenth century, when technological progress spread to pre viously unaffected industries.

The Industrial Revolution is usually dated between about 1760 and 1830. Britain is usually thought of as the location where it all happened, but a large part of the new technology was the result of work done in other European countries and later in the United States. It is not easy to generalize about the kind of technological change that occurred. Some have proposed that the main feature was the substitution of inorganic for organic materials (Wrigley, 1987). Another tries to define the Industrial Revolution as an increase in energy inputs, especially inanimate energy, and focus on steam power as the most significant advance (Cipolla, 1965a). Others such as Paulinyi (1986) focus on the use of machines instead of hand tools. Yet these generalizations fail to do justice to the rich diversity of progress in these years. The growth of cotton at the expense of wool and linen, the improvements in the efficiency of water power, the development of gaslighting, the advances in the machine tool industry, and the invention of food canning, to mention just a few examples, really share few characteristics save their capability to increase both the quantity and quality of the supplies of goods and services. As McCloskey (1981, p. 118) put it, the Industrial Revolution was not the Age of Cotton, or the Age of Steam; it was an age of improvement. Yet improvement was not pervasive. Large sectors of the economy, employing the majority of the labor force and accounting for at least half of Gross National Product were, for all practical purposes, unaffected by innovation before the middle of the nineteenth century. Services, construction, food processing, and apparel making were some of the sectors which in 1850 used techniques unaltered for centuries. The reason why some industries changed and others did not has little to do with either the demand side or the supply of raw materials and coal; it simply reflects that technological opportunities and constraints determined the outcome of a process driven by the supply side of technology.

It is useful to divide the technological changes during the Industrial Revolution into four main groups: power technology, metallurgical, textiles, and a "miscellaneous" category of a variety of industries and services.

Power technology. The protestations of some economic historians notwithstanding, the steam engine is still widely regarded as the quintessential invention of the Industrial Revolution. Its background was not purely British; it is better to think of it as the result of an international joint effort. The basic idea for the construction of an atmospheric engine was based on the realization of the existence of an atmosphere. What seems today a commonplace insight, was the fruit of the work of Evangelista Torricelli, a student of Galileo's, and Otto von Guericke, the major of Magdeburg, famous for his experiment in which two teams of horses could not separate two hemispheres enclosing a vacuum.[42] The existence of an

[42] The progress from the discovery of the atmosphere to the construction of atmospheric engines is sometimes regarded as an example of how science influenced technology at this early stage. The relation was, however, fully reciprocal. The discovery of the atmosphere was prompted by the puzzling observation that vacuum pumps could not raise water more than 32 feet. Galileo was one of the people who struggled with this, and Torricelli was probably influenced by his Master's interest in the problem.

atmosphere and its pressure may well have been known to the Chinese, but what followed in the second half of the seventeenth century was a typical European story. It occurred to many who had grasped the newly discovered phenomenon that if one could create a vacuum repeatedly, the force of atmospheric pressure could yield a novel source of power. The Marquis of Worcester, among others, suggested in 1663 a machine utilizing condensation for this purpose. The first known model was built in 1691 by Denis Papin, an assistant to Huygens, who showed in a prototype how a piston could be moved up and down a cylinder using steam. Application to a useful purpose followed suit: Thomas Savery built a first working steam engine in 1698, though this device was really a suction pump, which condensed steam in a closed vessel and sucked up the water by the vacuum. A different version was perfected in 1712 by Thomas Newcomen, who, unlike Savery, used atmospheric pressure in a machine that was alternately heated and cooled, so as to create repeated vacua by condensation. Newcomen's engine was a far more complex and sophisticated machine than Papin's prototype, yet it was within the ability of the craftsmen of the time and it was safe. It was powerful enough to pump water out of mines, and despite the awkward dimensions of the machine, its voracious appetite for fuel, and the difficulty early eighteenth century mechanics had in achieving hermetic sealing, the Newcomen machine was widely employed. Within a few years of its inception, the machine spread to France, Germany, and Belgium, and by 1730 it was operating in Spain, Hungary, and Sweden and later in the American Colonies. Newcomen's machine solved drainage problems in the Cornish tin mines, as well as in the deep coal mines in the north of England. But above all, it was the first economically useful transformation of thermal energy (heat) into kinetic energy (work).[43]

Yet to the nonspecialist, the steam engine will forever remain associated with the name of James Watt. The basic improvement that Watt introduced, was to separate the condenser from the

[43] Strictly speaking, fire arms were also based on that principle. The cannon is the first known example of a controlled conversion of thermal energy to kinetic energy; it is a one-cylinder internal combustion machine. Yet it took another half-millennium to perfect the internal combustion engine, and its debt to firearms seems dubious.

piston-cylinder, so that the latter could be kept constantly hot. This separation greatly reduced the fuel requirements of the machine and permitted it to be used almost anywhere. Watt's ingenuity provided many further improvements on the steam engine, including steam-jacketing to keep the cylinder hot; a transmission mechanism know as the "sun-and-planets" gears, which converted the reciprocating motion of the atmospheric engine to the rotative motion needed in textile mills and other industrial applications; and a parallel motion gear that allowed steam to be introduced alternately into both ends of the cylinder, thus creating a double acting engine that utilized the push as well as the pull of the end of its beam. The double-acting expansion machine used the steam above the piston to drive it down but cut off the steam after the piston has moved part of the way in order to save fuel. As a result, fuel efficiency was raised from less than 1 percent in the Newcomen engines to around 4.5 percent in Watt's designs. The principle of double-acting engines had been known in Europe since the fifteenth century, and used in pumps and bellows, but its application to the steam engine was by far the most successful (Reti (1970)). The utilization of the expansive power steam was regarded by Usher (1954, p. 354) as the difference between an atmospheric and a genuine steam engine. Equally important was the "governor" for regulating the speed of engines that foreshadowed twentieth century feed-back servo-mechanisms, which form the basis for cybernetics. The device consisted of two balls that were pushed outward when the speed of operation increased and that lifted an arm attached to the valve in the steampipe. Similar automatic regulators had been applied earlier to windmills, but Watt's invention symbolizes the desire for full control and automation that increasingly permeated the technique of the time.

Watt's work was paradigmatic of the kind of mind that helped make the Industrial Revolution: a combination of inventive genius and a desire to cut costs, minimize wear and tear, and extract "the last drop of 'duty' from the last puff of steam in his engine" (Cardwell (1972, p. 93)). Watt himself, in his oddly-written third person autobiography wrote that "his mind ran upon making engines *cheap* as well as *good*." The search for economic value in technology in addition to functionality and beauty is the culmination of a millennium of European technological rationality. Yet rationality meant nothing without technical ability, and Watt's mechanical talents bordered on the virtuoso. In short, James Watt is, in the history of power technology, a figure comparable to, say, Pasteur in biology, Newton in physics, and Beethoven in music. Some individuals did matter. We should bear in mind, however, that he stood on the shoulders not only of Papin and Newcomen, but also on those of John Wilkinson, whose new boring machines supplied Bouton & Watt with cylinders of great accuracy, and on those of his partner Boulton, with whom Watt formed a classic inventor-entrepreneur couple (Scherer (1984), pp. 8–31). The steam engine became a familiar sight in eighteenth century Britain. It is now reckoned that close to 2,500 engines were built in the eighteenth century, of which about 30 percent were made by Watt, though he was by far the largest producer (Kanefsky and Robey (1980)). The most important single user of steam power were mines, with 828 engines in collieries and another 209 in copper and lead mines.

Watt's patent expired in 1800, and a new genius applied himself to the construction of a revolutionary steam engine. Watt had felt that a high-pressure machine—already suggested by Savery—was too dangerous to be practical. But in 1802 Richard Trevithick built a machine that developed a pressure of ten times the atmosphere. These high pressure machines were smaller in size and more economical. This made it possible to place powerful engines on boats and horseless carriages. A steamboat prototype was built by the Marquis de Jouffroy in France in 1783 and by John Fitch and James Rumsay in the United States in 1787, and made practical by Robert Fulton in 1807. Within ten years Fulton's boats were dominated by high-pressure engines, pioneered in the United States by Oliver Evans (Hindle (1981), p. 55). Nonetheless, the Watt stationary engine also continued to improve, and the two types thus existed side by side throughout the nineteenth century.

The next step in the development of steam power was compounding, that is, the use of the same steam in more than one cylinder, one of which is high pressure. In these engines, after the steam does its duty in the high-pressure cylinder, it is admitted into a larger cylinder where it drives down a piston using the expansive principle.[44] The first successful application of compound engines was by Arthur Woolf in 1803.[45] Compounding became practical only in 1845, when John McNaught (after whom the verb "Mc-naughting" was coined) perfected the process. The advantage of compounding was fuel saving: Woolf's compound engine raised fuel efficiency to 7.5 percent (compared to 4.5 percent in Watt's engines), and the sophisticated Corliss compound steam engine of 1878 had a thermal efficiency of over 17 percent.

The success of the steam engine preceded the establishment of a science that formalized the principles upon which it was based. In 1824, the Frenchman Sadi Carnot, upon observing a working steam engine and asking himself why high pressure engines were superior, developed the kernel of what was later to be known as thermo-dynamics. Carnot also maintained that "to take away England's steam engines today would amount to robbing her of her iron and coal, to drying up her sources of wealth, to ruining her means of prosperity and destroying her great power" (cited by Cardwell (1972), p. 130). Economic historians would probably disagree. The assessment of the contribution of the steam engine should not be based on the gross achievements of the steam engines, but on the marginal contribution of steam relative to its next best alternative.

That next best alternative was water power, still an important source of power in the British economy by 1830, and the dominant source of energy in Switzerland and New England. The gains that the steam engine provided relative to water power were fairly small (Von Tunzelmann, (1978)). But this fact should not distract from the achievement of the men who made the Industrial Revolution. Indeed, it confirms it. Between 1750 and 1850, fundamental improvements were introduced in the utilization of water power. Improvements in the efficiency of water power use explain the slow

[44] The original inventor of the compound steam engine was Jonathan Hornblower who patented a compound engine in 1781. His invention was determined to be an infringement on Watt's patent, however, and compounding was made to wait for another two decades.

[45] Woolf based his compound engine on a scientific "law" which, he claimed, showed a linear relation between excess pressure and the increase in the volume of steam when allowed to expand to atmospheric pressure. This erroneous theory, quite at variance with Boyle's Law, was the basis for a sound mechanical design. Cardwell remarks, quite correctly, that "not for the first time in history a useful advance was made on the basis of a defective theory" (Cardwell, 1971, p. 155).

diffusion of steam power in many places. The improvements in water power were associated with constant improvements in the understanding of the theory of hydraulics. The most important advance was the breastwheel, which was introduced by John Smeaton in the 1750s and soon spread all over Britain. Breast-wheels receive water at an intermediate point between the summit and the bottom of the wheel, and thus in a sense are a compromise between overshot and undershot wheels. The breastwheel was as efficient as the overshot wheel, but had the advantage that the wheel turned in the same direction as the flow of water in the tail race, which allowed it to work under flooding conditions. Smeaton was a practical engineer, not a scientist, and was known for his distrust of scientific theory. This distrust was not altogether mis-placed: it was not until the 1750s that scientists realized that different principles applied to the gravity driven wheels of overshot mills and the impulse driven wheels of undershot mills. For a long time, practical men without formal training in hydraulics kept making the important improvements. Smeaton's work was perfected by the introduction of the sliding hatch, introduced by John Rennie in the 1780s, which allowed the breastwheels to adapt to varying water levels. Waterwheels were increasingly made of iron parts, which reduced wear and tear.

Across the channel, French engineers were equally successful. The French engineer Jean Charles Borda was the first to attack the theoretical problems of water power in 1766, though his work was too abstract to be of immediate use and not recognized until after 1810 (Reynolds, 1979). Jean Victor Poncelet used Borda's ideas to modify the undershot waterwheel to build the famous Poncelet waterwheel, which used curved blades and fed the water through an inclined hatch (1823). The so-called "column of water" engine, extensively employed on the Continent, combined ingeniously the idea of a Newcomen engine with the pressure of water (rather than the atmosphere). Water power technology was further advanced by the invention of the water turbine. The idea originated in the eighteenth century with Leonhard Euler, the Swiss mathematician, who showed (together with his son Johann) that by using the force of water coming *out* of the vanes of a waterwheel, the entire energy of the flow could be converted into useful work. The difference between this concept and the waterwheel is that in the latter the water does not move relative to the buckets or vanes, while in the former the power is derived from the water flowing rapidly through curved passages driving the engine. The practical implementation of the idea took many years of tedious work by many engineers (mostly in France), culminating in the success of the Frenchman Benoît Fourneyron in 1837. The turbine was introduced into the New England textile industry and "delayed for decades the dominance of steam power in that industry" (Rae (1967), p. 338). By this time, advances in water power technology had been so impressive that one historian notes that engineers had in theory, and to a large amount in practice, complete command over water power (Cardwell (1971), p. 184). Nonetheless, the utilization of water power remained constrained, not so much by the water mills themselves as the lack of a scientific understanding of watershed hydrology and the requisite data on rainfall (Gordon (1985)).

The Industrial Revolution was an age of power technology, and the prospects must have seemed unlimited. The economic impact of the steam engine during the Industrial Revolution may at first not have been as large as Carnot thought, but in the second half of the nineteenth century steam power penetrated every aspect of economic life in the entire Western world and beyond. Power technology, in conjunction with other inventions, created the gap between Europe and the rest of the world, a temporary disequilibrium that allowed the Europeans to establish global political and military domination.

Metallurgy. Prior to the Industrial Revolution, metallurgy had always been an empirical and experimental art, in which gifted amateurs and semi-professionals tried to solve complex chemical and physical problems that baffled those engaged in the day-to-day manufacture of metals. Their success is a monument to their determination to produce materials that satisfied their needs better and more cheaply. Among these materials, iron ruled supreme. There were no substitutes for it in durability, versatility, and malleability. We have seen that the later Middle Ages witnessed the production of cast iron, which has a relatively high carbon content and a low melting point. It was brittle and hard, and could therefore not be shaped using the traditional blacksmith's tools, but it could be cast to make pots, ovens, and cannon. Most products, including machine parts, nails, locks, and tools needed to be shaped in forges, and for these wrought iron was needed. The process of turning pig iron, the output of blast furnaces, into wrought iron remained a major bottleneck in the metal industry. During the eighteenth century an extensive search was conducted into this problem in Britain. The Wood brothers pioneered the so-called potting process using crucibles or "pots" to heat the pig iron (Hyde (1977), pp. 83–88). The problem's ultimate solution by Henry Cort in 1784 was a skillful combination of a number of elements, such as the reverberatory oven that had been used in glass furnaces for centuries and the rolling of heated metal using grooved rollers. Cort's puddling and rolling process was typical of many great inventions of the Industrial Revolution: the culmination of a dispersed and drawn out search for the solution of a difficult but economically important problem. Cort's process, after several further improvements in the late 1780s, took the world of metallurgy by storm. The small independent forge, until then the source of all wrought iron, vanished, to be replaced by the larger puddling furnaces. The supply of high quality and cheap wrought iron grew dramatically, making iron almost literally the building block of the Industrial Revolution.

The eighteenth century also witnessed another invention comparable in fame: the use of coke in blast furnaces. Coke is purified bituminous coal, and its use in smelting was pioneered by Abraham Darby. Coke had been employed earlier in industry, but its use in blast furnaces dates from about 1709. Much nonsense has been written about coke smelting, including the famous myth that it was triggered by a scarcity of wood, or that the diffusion of the invention was impeded by the secrecy of the ironmaster's family. The simple problem was that for a long time coked iron contained silicone, which made it more costly to convert coked pig iron into wrought iron (Tylecote (1976), pp. 108–09). The growth of coking after 1750 has been explained by further technological progress, especially Darby's son's success in remelting the pig iron in a so-called "foundry furnace," which removed the silicone. Other improvements included the making of cokes in closed ovens shaped like beehives, and by the replacement of old-fashioned bellows by new water-powered blowing cylinders, invented around 1760 by the versatile John Smeaton. Between 1760 and 1790, coke replaced charcoal in British iron smelting and was gradually introduced on the Continent. Blast furnaces became bigger and more efficient, producing better-quality iron at lower prices. Another major breakthrough was achieved by the Scotsman James Neilson, who hit upon the idea of utilizing the blast furnace's own gases to preheat the air inside (1829). This procedure simultaneously reduced fuel requirements by a factor of three and created temperaures hot enough to use fuels other than coked bituminous coal, such as anthracite. It was this process that made it possible to exploit the "black-band" iron ore deposits in Scotland and allowed Scotland to compete with the Black Country and Wales in pig iron production.

The product that most stubbornly resisted innovation was steel. Chemically, steel is an intermediate product, halfway between the almost carbonless wrought iron and high-carbon pig iron. Steel can be made from iron by adding carbon to low-carbon wrought iron (cementation or carburization); by removing carbon from high-carbon cast iron (decarburization); or by mixing high and low carbon scraps of iron together (cofusion). A fourth process produced steel directly from ore by packing it in crucibles with pieces of special wood. This last process was known as "wootz" steel, first produced in India (in Hyderabad). Steel's desirable physical properties (resilience, tenacity, and flexibility) made it ideal for razors, weapons, shears, springs, and machine parts, but it was prohibitively expensive. In the West, steel was produced by cementation, a process known first in Asia Minor around 1000 B.C. The production of this "blister steel" entailed "baking" the wrought iron by heating it in direct contact with charcoal and hammering it for extensive periods to spread the carbon through the metal. The blast furnace offered new opportunities to make steel, especially because ores high in manganese allowed a better control of the carbon-removal, allowing some residual of carbon to be left. By the seventeenth century, Europeans had learned that steel could be improved by remelting and hammering small pieces of it at very high temperatures, thus spreading the carbon somewhat more evenly. The production of high quality steel was perfected in about 1740 by Benjamin Huntsman, who used coke and reverberatory ovens to generate sufficiently high temperatures to enable him to heat blister steel to its melting point. This way he produced a crucible or cast steel that was soon in high demand by instrument- and clockmakers.

The importance of crucible steel was that it could be cast, and eventually the production of larger ingots became possible, through the coordinated operation of many crucibles. The German steel manufacturer Alfred Krupp pioneered these techniques in the casting of steel cannon. His sixpounder was one of the great sensations of the Crystal Palace exhibition of 1851, as was a gigantic steel casting weighing 4,300 pounds. Yet steel remained too expensive to be of widespread use during the critical years of the Industrial Revolution. Wrought iron rather than steel was the main material until 1860.

Although metallurgy was an art rather than a science, and *all* the advances described above were the result of tinkering and trial and error searches, it is not quite true that the absence of a science of chemistry meant that the ironmasters had no idea what they were doing.[46] The Frenchman René Réaumur published a famous book on metallurgy in 1722 in which he fully elaborated the fundamental fact of metallurgy that the qualities of iron are largely determined by its carbon content. This insight was later corroborated by the Swede T. O. Bergmann in 1750.

Textiles. In the textile industry as in metallurgy and power technology, technological progress was the result of the joint and cumulative efforts of many individuals. A typical innovator in those years was a dexterous and mechanically inclined person who became aware of a technical problem to be solved and guessed approximately how to go about solving it. The successful inventors were those who put the pieces together better than their colleagues, or those who managed to resolve one final stubborn difficulty blocking the realization of a new technique.

The central technical problem in textiles was that of spinning. Since time immemorial, the crucial operating part in the spinning process had been the human finger, the thumbs and pointers of millions of women who gave the raw material in the rovings the "twist" that made it into yarn. The spinning wheel increased the efficiency of the spinner's work, but did not replace the human

[46] The pathbreaking work of Lavoisier and his colleagues establishing modern chemistry was carried out in the 1780s, practically simultaneously with the development of the puddling and rolling process.

finger as the tool that transformed the material. The search for a replacement for the human hand in cotton spinning was taken up by an Englishman named Lewis Paul, who pioneered the idea of using rollers replacing the fingers in drafting out the fibers. His patent was taken out in 1738, but it is Richard Arkwright to whom the mechanization of spinning is usually credited, more than thirty years later. Arkwright's machine, the "throstle" or "water frame" differed from Paul's in one respect: he used *two* pairs of rollers moving at different speeds and separated by a distance about equal to the length of the longest fiber to be spun. The net result was that Arkwright's machine worked, whereas Paul's did not. Yet the water frame was incapable of spinning the finer yarns, as these would have snapped when they were wound on the bobbins.

The water frame was complemented by another invention, the spinning jenny, which was patented a year after the throstle, but actually invented in 1764. Its inventor, James Hargreaves, reputedly hit upon the idea after watching a spinning wheel fall on its side and continuing to spin for a few more seconds. He realized that it was possible to "draft against the twist," that is, to impart the twist not by the movement of the fingers but by the correct turning of the wheel itself. The jenny twisted the yarn by rotating spindles that pulled the rovings from their bobbins, with metal bars playing the role of human fingers guiding the spun yarn onto the spindles by means of a "faller wire." It allowed a large number of threads to be spun at the same time, but the quality of the yarn was rather uneven and it was suitable only for weft. Moreover, the jenny was an extremely uncomfortable machine to work with, forcing adult spinners practically to bend double (Hills, 1979). Combining the throstle's rollers with the multiple spindles of the jenny led to the "mule," the ultimate spinning machine (patented by Samuel Crompton, 1779). The mule, one of the most famous inventions of all times, consisted of a carriage which was driven back and forth. In so doing, the spindles mounted on it turned quickly and together with the rollers imparted the twist on the yarn which could then be wound on bobbins. At no stage was the yarn subject to much strain, and thus the chances for breaking were much reduced. Consequently, not only could the mule make a much cheaper cotton yarn, but it could also make a yarn of unprecedented fineness, strength and uniformity. As a result, cotton became a growth industry the like of which no one had ever seen. Until Crompton, the cotton yarn spun in England was not strong enough to serve as warp and so cotton was used in combination with other yarns. The mule made the all-cotton cloth possible. It was especially suitable for finer yarn; coarse yarns continued to be spun by jennies for a long time, as they were cheaper and could be readily used in domestic industry.

Modifications in spinning were subsequently introduced, but the main breakthroughs had been achieved by 1780. Application of steam power to the new machine followed in the 1780s, although animal- and water power dominated in the early years. Some idea of the magnitude of the improvements attained can be gained from Chapman's (1972) calculations of the number of hours needed to spin 100 lbs of cotton. The "old" technology was the Indian hand spinner, who took about 50,000 hours. Arkwright's rollers and the mule brought that number down to around 300 hours in the 1790s, and the automatic "self-actor" patented by Roberts in 1830 reduced it to 135. The self-acting mule was a triumph of British engineering, in the words of Mann (1958, p. 290) "an almost perfect machine." Yet despite its obvious advantages, it was adopted so slowly that the patent authorities decided to extend the patent by another seven years. The main reason seems to have been that self-actors were expensive, and sources of long-term credit for fixed capital were scarce. Moreover, the self-actors were better suited for coarse (low count) yarns until about 1860.

Although the improvements in spinning were the most spectacular, other stages in cotton production did not lag far behind. The cotton gin, invented by Eli Whitney in 1793, assured the supply of cheap raw cotton to Britain's mills. A carding machine was patented by Lewis Paul in 1742 and later improved by Arkwright, who pioneered the use of large rollers to prepare the rovings for the water frame. The finished yarn was bleached using chlorine, a process invented by Claude Berthollet in 1784 and improved by Charles Tennant in 1799 by combining chlorine with slaked lime to make bleaching powder. In 1783 Thomas Bell invented metal printing cylinders that printed patterns on the finished cloth. In weaving, the introduction of machines was slower. Improvements at first took place largely in the equipment used in domestic industry, which explains the long prosperity and survival of the domestic hand loom weavers. The dandy loom, patented by Thomas Johnson in 1805, moved the cloth beam automatically, speeding up the weaving process with no extra effort. The first power loom was built in 1785 by Edmund Cartwright, an enterprising and idealistic clergyman, but the machine did not work properly until about 1815, and the finer yarns (which were more subject to breakage) were not woven by power looms until the 1830s.

Thus, during the brief period between 1760 and 1800, a feverish inventive wave focused on the manufacturing of cotton. Cotton combined qualities that are attractive to both consumers and producers: it takes dyes well, launders easily, and ventilates much better than linen and wool. Its fibers lent themselves easily to mechanization compared to its main competitors, wool and linen, which were relatively slow to follow cotton. The supply of the raw material was elastic. No wonder, then, that cotton grew at a rate never witnessed before in textiles, and is still regarded as the quintessential growth industry during the early stages of the Industrial Revolution.

The English woolen industry had consisted since the sixteenth century of two branches: the woolen and the worsted industries. In worsted, the preliminary combing of the wool was the slowest to mechanize. Although a number of patents were taken out on combing machines, successful combing machines were not available before 1827 and not perfected until Josué Heilmann, a Frenchman, patented his combing machine in 1845. On the other hand, Arkwright's double-roller device was well-suited to the spinning of worsted yarns. In wool the order of mechanization was reversed: preparation was mechanized before spinning. Water-driven carding machines were operating in Yorkshire in the 1770s, whereas the adaptation of the mule to the spinning of wool was not solved satisfactorily until 1816. In weaving, power looms were applied to worsteds after 1820, but the diffusion of these machines was slower than in cotton. In wool, the yarns were too fragile for the power looms, and mechanization did not occur until the 1840s. Hand loom weavers in the woolen industry survived longer than in cotton or worsteds. In the finishing processes of the woolen industry, innovation encountered some resistance, but here too mechanization proved inevitable.

Worsteds, like silk, were a delicate and relatively up-market fabric woven into patterns using the so-called Jacquard loom, one of the most sophisticated technological breakthroughs of the time. The Jacquard loom was perfected by Joseph Marie Jacquard, a Lyons silk weaver, in 1801 after French inventors had been struggling for a century with the problem of a loom that could automatically weave patterns into fabric (Usher (1954), pp. 288–95). The patterns were coded on cards by means of holes representing the information in a binary code. The cards were probed by rods connected to wires, transmitting the information embedded in them.[47] Despite resistance among French weavers, the Jacquard loom spread rapidly. A decade after its invention, 11,000 such looms were operating in France. After 1820, its diffusion in Britain was rapid.[48]

In the ancient and venerable linen industry, mechanization turned out to be difficult. In flax, the filaments contain a rubbery substance that needs to be dissolved before spinning can take place. Here, too, determination and perseverance by ingenious pioneers eventually cracked the problem. The search was intensified during the Napoleonic Wars when the Continental textile industries were cut off from their sources of raw cotton supply. The Emperor Napoleon, who had a keen interest in industry, offered a large financial reward to whomever would do for linen what Arkwright and Crompton had done for cotton. The Frenchman Philippe de Girard perfected the idea of "wet spinning," soaking the raw flax in a hot alkaline solution before it went to the spindles. This idea was

[47] What became known as the Jacquard loom was in fact invented by two Frenchmen in the late 1720s. One of them, Basile Bouchon, was the son of an organ maker, which is significant because the organ is the first direct application of the binary coding of information. Bouchon's invention was improved in 1728 by Jean Baptiste Falcon. Bouchon coded his patterns on perforated paper, whereas the Falcon loom used separated perforated cardboard cards. Later, the great engineer and inventor Jacques de Vaucanson turned his attention to the loom (1775) and suggested using a cylinder with holes punched in it, and by attaching the warp threads directly to a hook manipulated by needles probing the perforated cylinder. Vaucanson's ingenious device was never put into operation. Jacquard's own contribution to the machine was to combine the best aspects of the Falcon and the Vaucanson devices and dispensing with the drawboy who traditionally helped in directing the pattern woven.

[48] The Jacquard loom inspired Charles Babbage in the design of his famous Analytical Engine as well as Richard Roberts' multiple spindle machine, which used a Jacquard-type control mechanism for drilling rivet holes in the wrought iron plates used on the Britannia tubular bridge (Rosenberg and Vincenti (1978), p. 39). It also inspired the American Herman Hollerith in the 1890s to use punch cards to store the 1890 U.S. Census information.

introduced into Leeds in 1825, and the mechanization of linen spinning spread rapidly. The preparatory stages of flax processing, such as the scutching and heckling of the material (roughly corresponding to carding in wool or cotton or combing in worsteds), which had been a highly labor-intensive cottage industry, were mechanized in the 1830s, though the finest yarns remained those heckled by hand. Moreover, the adaptation of the power loom to linen weaving ran against a barrier: the lack of elasticity caused the yarn to snap under strain. As late as 1850, there were only slightly over one thousand power looms weaving linen in the United Kingdom, as opposed to a quarter of a million in cotton and forty two thousand in wool and worsted. The technological difficulties in mechanizing the manufacture of linen cloth led to the sharp decline of this textile industry, with devastating effects on regions that had historically specialized in it, such as Ireland and western Belgium.

Miscellaneous technological progress. Which elements were responsible for making the Industrial Revolution possible in the late eighteenth century and not a century or two earlier? One factor surely must be the existence of a small but vital high-precision machine tool making industry. We have already mentioned the firm of John Wilkinson that made the cylinders used by Watt. It is only a mild exaggeration to say that Wilkinson and his colleagues made the difference between Watt and Trevithick on the one hand and Leonardo da Vinci on the other. Machine tools such as planing machines, milling machines, lathes, screw-cutting machines and so on permitted the creation of precise geometric metal forms, essential to machine making and uniformity. It was, in the words of one technological historian, "the most important step on the way to the production of machines by machines . . . it [became] possible to use iron and steel as the material whenever it appeared functional to do so" (Paulinyi (1986), p. 277).

Unlike textiles, the masters of the machine-tool industry were a closely-knit group, the members of which taught each other the secrets of the trade. The dynasty was started by Joseph Bramah, who had eighteen patents to his name, including an improved water closet, a wood-planing machine, sophisticated locks, and a spring-winding machine. In 1797, Bramah's foreman, Henry Maudslay, left the firm to start his own, devising a screw-cutting lathe that produced screws with unprecedented accuracy and at an affordable price. He built numerous machine tools used for sawing, boring, mortising, and so on (Woodbury (1972)). Maudslay in his turn trained three other toolmakers, all of whom made major contributions to the machine tool industry: Richard Roberts, James Nasmyth, and Joseph Whitworth. Nasmyth invented the steam-hammer, as well as milling and planing machines. Whitworth had no fewer than twenty three exhibits at the famous Crystal Palace exhibition of 1851, where the glories of British engineering were extolled. To his credit are, among others, a measuring machine that could measure up to one millionth of an inch and the standardization of screw threads. Finally, Richard Roberts was one of the most brilliant engineers of his time, making fundamental contributions to half a dozen inventions, including a multiple-spindle drilling machine controlled by a binary logic mechanism similar to the Jacquard loom. Robert's improvements to Cartwright's power loom turned it from a curiosum invented by a well-meaning eccentric into the backbone of the British cotton industry (and the doom of Britain's hand loom weavers). His most famous invention was the self-actor (1830) which, as we have seen, wholly automated the mule-spinning process and turned the spinner from a skilled operator to little more than a machine tender. Other machine tool makers, not all associated directly with the Bramah "dynasty," deserve mention. Jesse Ramsden built a dividing machine for the accurate graduation of circles, crucial to the construction of navigational and surveying instruments (1763). Matthew Murray of Leeds and Bryan Donkin of London also contributed to the design of improved machine tools. The famous Portsmouth blockmaking machines, devised by Marc I. Brunel and Henry Maudslay around 1801 to produce wooden gears and pulleys for the British Navy, were automatic and in their close coordination and fine division of labor, resembled a modern mass-production process (Cooper (1984)). William Fairbairn, a producer of improved waterwheels, steamboats, and boilers, was a worthy heir to John Smeaton. In the United States, Eli Whitney, John Hall, Simeon North, Thomas Blanchard, and others pioneered new machine tools, paving the way eventually for the "American System" of interchangeable parts.

In the ceramic and glass industries, the location and timing of technological change were quite different than in textiles and metallurgy. Here the Continent played a major role, and most technical developments occurred before 1750. In the seventeenth century Dutch craftsmen, struggling to imitate Chinese blue-and-white procelain, hit upon the processes of making the tin-and-enamel wares for which Delft is famous. Porcelain was produced successfully in 1708 by J. F. Böttger, who was employed by August the Strong, Prince-Elector of Saxony. Dresden chinaware was made in Meissen, and kept strictly secret by the Saxons. It took the rest of the Continent a few decades until they were able to copy the process. England came up with a close substitute: the salt-glazed earthenwares produced by John Astbury of Staffordshire in about 1720. By 1759, when the renowned Josiah Wedgwood started his factory in Burslem, he could draw upon a venerable tradition of making an excellent product. In the preceding century, the British pottery industry had adopted coal as a fuel for most purposes, as well as new raw materials such a ball-clay and flint. Wedgwood is justly famous for his rationalization of production employing steam engines, a fine division of labor, and his ability to market a relatively inexpensive product to snobbish consumers. Wedgwood competed successfully with the more expensive hard-paste porcelains made on the Continent.

In the glass industry, too, Britain followed the Continent. In or around 1688, French glassblowers adopted the cast-plate process, which produced flat glass (by far the most important output of the industry) with a much more even and flatter surface than the cheaper method of blown "crown" glass. The great Royal Manufacture at St. Gobain led in the production of window glass and mirrors using this system. Its adoption in Britain dates only from the 1770s, and even then it remained dependent on French know-how. On the other hand, the British pioneered glassmaking using a coal-fired reverberatory furnace. Through covered crucibles, the smoke was prevented from discoloring the glass, a technique the French did not master until the Revolution (Scoville (1950), p. 43). Britain had no monopoly on invention, but when it was behind, it shamelessly borrowed, imitated, and stole other nations' technological knowledge. Another good example of Britain applying new technology developed elsewhere was chemicals. Berthollet, the inventor of chlorine bleaching, was French, as was Nicholas Leblanc who developed the soda-making process named after him. Leblanc reacted salt and sulphuric acid to produce sodium sulphate, which after heating with lime or charcoal yielded raw soda together with hydrochloric acid, a noxious by-product. The Leblanc process became the basis of the modern chemical industry and is regarded as one of the most important inventions of the time.[49]

In the adoption of soda, Britain was relatively slow, and only in the 1820s did it start to adopt Leblanc's process on a large scale. The explanation usually given for this delay is the high tax on salt, which made artificial soda more expensive than vegetable alkali. Once the salt tax was repealed, British soda production grew rapidly and by the 1850s exceeded French output by a factor of three (Haber (1958), pp. 10–14). In other areas of the chemical industry, however, the British led the Continent from the start.[50] Only after the middle of the nineteenth century did the Germans establish firm leadership in chemistry due to their superior professional chemists.

An interesting example of international collaboration in technological progress was the development of gaslighting. Although this invention was probably modest in its contribution to national income, it was quite important in its effect on the quality of life: gas-lit streets were safer, gas-lit homes promoted literacy, gas-lit theatres made entertainment more sophisticated, and gas-lit factories made night work cheaper and more efficient. Prior to 1780, lighting technology had changed little from antiquity. Lamps burning rapeseed and similar oils provided a smoky light. Candles were mostly made of tallow and produced a smoky and malodorous

---

[49] Leblanc's story is an illustration of the risks to which an inventor is subject. The French academy had offered 100,000 francs for the person who could produce soda from sea salt. Leblanc solved the problem (1787) but did not receive the prize. He did, however, secure the help of the Duke of Orleans (his employer) and in 1789 started a joint venture in St. Denis. As it happened, that year was a bad time to start a commercial enterprise in partnership with a prominent member of the *ancien régime*. The revolutionary government annulled Leblanc's patent under the pretext that soda manufacturing was a matter of national security. The price of soda fell precipitously, and after Leblanc failed in his desperate attempts to get some compensation for his invention, he committed suicide in 1806. Only in 1855 did his heirs receive a financial honorarium for his invention.

[50] In 1746, John Roebuck perfected his lead-chamber process for the production of "vitriol" (sulphuric acid), of importance in bleaching and metallurgy. James Keir, a glass manufacturer, was one of the leading applied chemists of his time.

flame. Wax candles were superior, but too expensive to be used widely.

Gas lighting was a joint German-Anglo-French project. The use of gas as a source of light was first pointed out by the Belgian physicist Jean Pierre Minkelers and the German pharmacist J. G. Pickel in the late 1780s. In 1799, the Frenchman Philippe Lebon combined a wood-derived gas with the Argand burner, developed by the Frenchman Aimé Argand in the early 1780s.[51] In Lebon's so-called "thermolamp," the gas and the air were introduced separately, and the heavier by-products were collected in a special receptacle. In 1798, the Scotchman William Murdock, who worked for Boulton & Watt (which had bought Argand's patent in Britain), lit his forge with gas derived from coal, which soon proved superior to wood as a source. By 1807, Manchester's cotton mills and London's Pall Mall were illuminated by gas. The Englishman Samuel Clegg showed how gas could be distributed from a central production site to individual consumers using hydraulic gas-mains. His son-in-law, John Malam, developed a metering system that could monitor the quantities used by each customer (1819). Following the return of peace in 1815, the idea spread rapidly through the Continent: the work of a German chemist appropriately named Wilhelm Lampadius led to the illumination of Berlin streets in 1826. Further innovations, such as filtering the gas through quicklime to remove the noisome odor caused by sulphurite hydrogen, helped to reduce the price and improve the quality of gas lighting.

One of the greatest macroinventions of all times occurred during the heyday of the Industrial Revolution, yet is rarely mentioned in connection with it, namely the invention of ballooning by the Montgolfier brothers in France in 1783. Surely, in terms of novelty and originality alone, ballooning is a truly epochal invention. Since time immemorial, people had dreamed of human flight, but some ill-fated attempts aside, it had never happened. The few attempts made to fly were typically in imitation of birds, trying to equip the flyer with wings and a tail without the faintest under-

[51] Like his compatriot Leblanc, Argand failed to capitalize on his burner because of his inability to prevent widespread imitation. He died impoverished and bitter in London in 1803.

standing of the principles of aeronautics. Montgolfier was the first to take an entirely different direction of thought, and one that proved to be successful to a degree and a speed that is little short of amazing. Joseph de Montgolfier was aware of Henry Cavendish's discovery of a gas lighter than air, hydrogen, in 1766. He believed that fire gave off a similar gas, and that when captured in a closed vessel, that vessel would be lighter than air and thus rise. The reasoning was thus partly fallacious. What made the Montgolfier balloons rise was not a gas lighter than air but air itself which, heated, expanded and thus reduced its specific weight. Nevertheless, success was immediate. Within half a year following the emergence of the idea, the famous demonstration in Annonay took place. On November 21, 1783, the first two human beings lifted off and travelled through the air and lived to tell about it. Except for a few military applications, ballooning was of course of limited direct economic effect until the Wright brothers returned to the idea of wings and tail a hundred and twenty years later. But there can be little doubt that the consequences of the invention of ballooning were far-reaching. Few things were more effective in accustoming people to technological success and alerting them to the ability of human ingenuity and creativity to control the forces of nature and do things never done before.

In mining, the years of the Industrial Revolution showed little progress. Coal became of ever-growing importance to the economy, in part because of its use in engines and metallurgy, and in part because a growing and more prosperous population demanded it for home heating. Yet aside from the use of steam engines in pumping water from the mines, the only major advance in coalmining was the safety lamp, invented by Humphrey Davy in 1815. By 1830 steam-driven fans came into use, reducing the dangers of explosion. The technical problems involved in using power machinery *inside* a mine could not be solved, however. The increase in production here consisted primarily of a shift *along* the supply curve, rather than an outward movement of the supply curves, as in textiles and metal.

The technological developments in the British manufacturing sector increased output by hitherto unimagined factors. The price of cotton cloth declined by 85 percent between 1780 and 1850. As elementary economic analysis shows, in a competitive economy technological progress in existing goods is usually transmitted to consumers through lower prices. The role of demand in the process was largely passive, as consumers slid down their demand curves.[52] Technological progress also created totally new goods and old goods at much better quality, but here too the shift in supply either satisfied an already existing demand or created new needs previously not explicity felt, invention being the mother of necessity rather than the reverse. It thus makes little sense to try to explain either the timing or the location of the Industrial Revolution by exogenous changes in consumer demand for goods and services.

Why did the many breakthroughs described above not occur earlier? Although the innovations that made the Industrial Revolution typically did not depend on new scientific knowledge, the fact remains that the technical problems that the engineers of the Industrial Revolution solved in metallurgy, power technology, and textiles were difficult. Given the tools, materials, and resources at the disposal of the most talented men in Europe, it is not surprising that it took a long time to tackle many of the challenges. Those who claim with Eric J. Hobsbawm that there was nothing inherent in the new technology of the Industrial Revolution that could not have developed 150 years earlier confuse scientific knowledge with technological ability. The problems, easy as they may seem to us perhaps today, were often simply hard. Even when all that was required was to combine previously known pieces of technical know-how into a gadget that would actually work, the effort put in by the inventor was great. The "invention" and "development" stages of technological progress were not yet separate. An idea by itself was not good enough, as a long line of frustrated inventors learned. Samuel Crompton, the yeoman-inventor of the mule, wrote that he spent "four and a half years at least wherein every moment of time and power of mind as well as expense which my other employment would permit were devoted to this one end." James Watt first conceived of his separate condenser in 1765; the first commercially successful machine began to operate only in 1776. The invention of the self-actor by Roberts involved a concentrated research effort (stimulated by the spinners' strike of 1824) that cost £12,000 and six years of hard work to complete. In addition to everything else, invention required people of unusual ability such as

[52] For more details see Mokyr (1985), esp. chs. 1, 2 and 4.

Watt, Smeaton, Trevithick, and Roberts as well as of unusual energy, daring, and luck like Arkwright and Boulton. The supply of talent is surely not completely exogenous; it responds to incentives and attitudes. The question that has to be confronted is why in some societies talent is unleashed at technical problems which eventually change the entire productive economy, whereas in others this kind of talent is either repressed or directed elsewhere.

Yet the story of the Industrial Revolution is far more than the tale of a handful of major inventors. The names mentioned above are but a small fraction of those who contributed to the new technology. Right below the "superstars" were hundreds and thousands of engineers, technicians, entrepreneurs, foremen, and gifted amateurs who made less spectacular contributions to technical progress. By adapting, modifying, improving, debugging, and extending earlier inventions, they were indispensable to the success of this uncoordinated, subconsciously joint project called the Industrial Revolution. In addition, we should not forget that there were even more numerous and more anonymous tinkerers and inventors who made things that did not work, or who were scooped by an earlier or luckier competitor. In an *ex post* sense these inventors were dispensable. But in an *ex ante* sense, all major and minor inventors operated in complete uncertainty about whether their project would succeed, and thus all were part and parcel of the Industrial Revolution.

6. THE LATER NINETEENTH CENTURY: 1830–1914

It is widely believed that before the middle of the nineteenth century, technological progress moved more or less independently of scientific progress, and that since then the interaction between science and technology has gradually become tighter. As we have seen, this view is only partially correct. The relation between scientific and technological progress after 1850 has confounded the best minds in the history of both fields (Mayr (1976)). Science and especially scientists were not totally irrelevant to technological change before 1850. Nor is it true that the imaginative, original, energetic, bold, but basically serendipitous, untrained, and unsystematic mind of the eighteenth century inventor disappeared after 1850. In fact, this kind of inventor is still quite important in the twentieth century.[53] As a general rule, it seems likely that in the past hundred and fifty years the majority of important inventions, from steel converters to cancer chemotherapy, from food canning to aspartane, have been used long before people understood *why* they worked, and thus *systematic* research in these areas was limited to ordered trial-and-error operations. The proportion of such inventions is declining, but it is still large today.

The function of science after 1850 was primarily to show what could *not* work rather than what *could*. The versatile inventor John Ericsson produced in 1853 a "caloric" engine based on regeneration of the heat produced. The machine was a failure because its inventor failed to realize a basic thermodynamic principle—that an engine cannot use the same heat over and over again, because the heat is converted into energy (Bryant (1973)). Even the best "tinkerer" inventors felt increasingly the need to collaborate with persons with a systematic training. Thomas A. Edison, for example, hired the mathematician Francis R. Upton who translated Edison's ideas into rigorous calculations. Inspired outsiders still played a crucial role in the invention process, and their openness of mind was often essential in the initial breakthrough. In the "development" stage of the process, however, scientific training and systematic work proved increasingly necessary. Inventors who wished to remain active had to become experts themselves or give up (Hounshell (1975)).

In the period after 1850, technological change differed increasingly from earlier periods in another aspect. Economies of scale were, of course, not new. Adam Smith had stressed the gains from the division of labor, and even the most casual observer knew that many types of machinery could not be made as cheaply in small sizes and doses, and that monitoring of workers, marketing, borrowing, and the absorption of new information were less costly for large than for small firms. The factory system emerged when cottage industry, in which the firm size had been constrained by the

[53] The classic defense of the individual inventor is in Jewkes, Sawers, and Stillerman (1969), who insist that "in many fields of knowledge, discovery is still a matter of scouting about on the surface of things where imagination and acute observation, supported only by simple technical aids, are likely to bring rich rewards" (p. 169).

size of the household, found it increasingly difficult to adopt new techniques. After the middle of the nineteenth century, these effects became more pronounced. To an ever-growing degree, learning by doing, large fixed costs in plant and equipment, positive externalities among different producers, network technologies, and purely technical factors such as the inherent scale economies in the chemical industry, all operated together in imparting declining average costs at the level of the industry as well as the firm. The relevance of this observation is that under conditions of increasing returns to scale the history of technology becomes a different tale. Brian Arthur (1985) has shown that under increasing returns, rational choice between competing techniques no longer guarantees that the most efficient technique will be chosen. In other words, the standard tools of economic analysis become inadequate in explaining the observed patterns of research and development, diffusion, and adoption. The more technology displays increasing returns, the more it is possible for an industry and even an economy to make a bad technical choice and to experience dire consequences of poor foresight.

The later nineteenth century is sometimes described as the age of steel and chemicals. It was that, but it was much more. In what follows, I can provide only a brief and selective survey of what must be regarded as a technological avalanche.

Steel. By 1850, the age of iron had become fully established. But for many uses, wrought iron was inferior to steel. The wear and tear on wrought iron machine parts and rails made them expensive in use, and for many uses, especially in machines and construction, wrought iron was insufficiently tenacious and elastic. The problem was not to make steel; the problem was to make *cheap* steel.

Henry Bessemer was a classic tinkerer, a professional inventor. His process of cheap steel followed a typical pattern: the inspiration came to him from the search for a different problem altogether (a high-speed spinning cannon shell). A sudden brilliant insight led to the basic steel conversion concept, but years of debugging were necessary until it was fully understood how and when the process worked, and why it failed in some cases. The Bessemer converter used the fact that the impurities in cast iron consisted mostly of carbon, and this carbon could be used as a fuel if air were blown through the molten metal. The interaction of the air's oxygen with the steel's carbon created an intense heat that kept the iron liquid, and thus by adding the correct amount of carbon or by stopping the blowing at the right time, the desired mixture of iron and carbon could be created. The high temperature and turbulence of the molten mass ensured an even mixture. As Bessemer soon discovered, phosphorus in the ores spoiled the quality of the steel, and thus the process was for a while confined to Swedish and Spanish ores. Cheap steel soon found many uses beyond the original spring and dagger demand; by 1880 buildings, ships, and railroad tracks were increasingly made out of steel.

Bessemer's invention occurred simultaneously with that of an American, William Kelly, who patented an identical process in June 1857, a year after Bessemer read his famous paper to the British Association for the Advancement of Science. Eventually Bessemer and Kelly came to an agreement to coordinate their patents. The almost uncanny simultaneity of the two inventions seems to lend support to the views of S. C. Gilfillan that individual inventors do not matter much, and that inventions will come along when the time is ripe. Whether such a conclusion is warranted is unclear. What happened in this case, and in many others, was that a "problem" was defined jointly by a perceived market need and by the state of the art as defined by previous inventions and accumulation of knowledge. This "problem" led ingenious men to explore various paths leading to a solution. Thus it came to pass that Bessemer and Kelly walked the same path and reached the finish line at almost the same time. A different path was taken by Continental metallurgists, who jointly developed the Siemens-Martin open hearth process, which used hot waste gases to preheat incoming fuel and air, and mixed cast iron with wrought iron in the correct proportions to obtain steel. The hearths were lined with special silica brick linings to maintain the high temperatures. The process allowed the use of scrap iron and low grade fuels, and thus turned out to be more profitable than the Bessemer process in the long run. Open hearth steel took longer to make than Bessemer steel, but its slowness was turned into an advantage because it permitted better quality control. Bessemer steel also tended to fracture inexplicably under pressure, a problem that was eventually traced to small nitrogen impurities. In 1900 Andrew Carnegie, the American steel king, declared that the open hearth process was the future of the industry. Like the Bessemer process, the Siemens-Martin process was unable to use the phosphorus-rich iron ores found widely on the European Continent. Scientists and professional metallurgists did their best to resolve this bottleneck, and it fell to two British amateur inventors, Percy C. Gilchrist and Sidney G. Thomas, to hit upon the solution in 1878. By adding to their firebricks limestone that combined with the phosphorus to create a basic slag, they neutralized the harmful phosphorus. It seems safe to say that the German steel industry could never had developed as it did without this invention. Not only were the cost advantages huge, but the Germans (who adopted the "basic" process immediately) also managed to convert the phosphoric slag into a useful fertilizer. While the Bessemer and Siemens-Martin processes produced bulk steel at rapidly falling prices, high-quality steel continued for a long time to be produced in Sheffield using the old crucible technique (Tweedale (1986)).

Chemicals. In chemistry, Germans took the lead. Although Britain still was capable of achieving the lucky occasional masterstroke that opened a new area, the patient, systematic search for solutions by people with formal scientific and technical training suited the German traditions better. In 1840 Justus von Liebig, a chemistry professor at Giessen, published his *Organic Chemistry in Its Applications to Agriculture and Physiology,* which explained the importance of fertilizers and advocated the application of chemicals in agriculture. Other famed German chemists such as Friedrich Wöhler, Robert Bunsen, Leopold Gmelin, August von Hofmann, and Friedrich Kekulé von Stradonitz, jointly created modern organic chemistry, without which the chemical industry of the second half of the nineteenth century would be unthinkable. And yet it was an Englishman, William Perkin, who made the first major discovery in what was to become the modern chemical industry.[54] The eighteen year old Perkin searched for a chemical process to produce artificial quinine, an antimalarial drug that was much in demand because Europeans were making deep inroads into tropical

[54] Perkin was trained by Von Hoffmann, who was teaching at the Royal College of Chemistry at the time, and his initial work was inspired and instigated by Hoffmann.

areas at about this time. While pursuing his work (1856), he accidentally discovered aniline purple or as it became known, mauveine, replacing the natural dye mauve.

German chemists then entered the search for other artificial dyes, and almost all further successes in this area were scored by them. In 1869, after years of hard work, a group of German chemists synthesized alizarin, the red dye previously produced from madder roots, beating Perkin to the patent office by one day. Despite the simultaneity of the invention, the discovery of alizarin in Britain marked the end of a series of brilliant but unsystematic inventions, whereas it was the beginning of a process in which the Germans established their hegemony in chemical discovery (Haber (1958), p. 83). In 1874 Perkin realized that he could not compete with the Germans, sold his factory, and devoted himself to pure research. The synthesis of indigo was carried out by Adolf Von Baeyer in 1880 after many years of hard work, but the costs of indigotin remained high. Here fortune favored the diligent: the accidental breaking of a thermometer in 1897 led *Badische Anilin und Soda Fabrik* (BASF) chemists to the use of mercury sulphate as a catalyst, and within a decade synthetic indigo replaced the natural product (Holmyard (1958b)). Another area in which British and German chemists first competed, but which was eventually dominated by Germans, was sulphuric acid. In 1875 W. S. Squire of England and Clemens Winkler at BASF developed the catalytic contact process for making sulphuric acid, but eventually the Germans came to dominate catalytic processes, which helped them become self-sufficient in ammonia, nitrates, and saltpeter in the twentieth century. The famous Haber process to make ammonia, developed by Fritz Haber in the early 1900s, and BASF chemist Carl Bosch's discovery of how to convert ammonia into nitric acid, made it possible for Germany to continue producing nitrates for explosives during World War I after its supplies of Chilean nitrates were cut off (Hohenberg (1967), pp. 29–30).

Progress in the chemical industry during the nineteenth century remained an international collaborative effort. In 1847, the Italian Ascanio Sobrero discovered nitroglycerine, the most powerful explosive hitherto known. Its instability caused numerous accidents, in one of which a Swedish manufacturer was killed. The victim's brother, Alfred Nobel, grimly set upon the task of controlling nitroglycerine, and discovered in 1866 that upon being mixed with diatomaceous earth nitroglycerine retained its full blasting power, yet could be detonated only using a detonating cap. Dynamite, as the new compound was called, was used in the construction of tunnels, roads, oil wells, and quarries. If ever there was a labor-saving invention, this was it.

In artificial fertilizer development, Britain remained in the lead. A Dublin physician, James Murray, showed in 1835 that superphosphates could be made by treating phosphate rocks with sulphuric acid. John Bennet Lawes carried out pathbreaking work in his famous experimental agricultural station at Rothamsted, where he put into practice the chemical insights provided by Liebig. In 1843 he established a superphosphates factory near London using mineral phosphates. Nitrogen fertilizers were produced from the *caliche* (natural sodium nitrate) mined in Chile. The third mineral crucial to plant growth was potassium, made from potash by burning wood. In 1870, forest-rich Canada was still the principal source of potash. By then, however, mineral deposits of potassium salts at Strassfurt in central Germany began to be exploited, the price of potash fell rapidly, and widespread application of this fertilizer began. By 1900 the Canadian asheries had disappeared (Miller (1980)).

In soda production, much work was done by engineers trying to improve the Leblanc process, which caused serious environmental pollution by spewing out clouds of hydrochloric acid gas and leaving a dark alkali ash known as "galligu." Technology dealt with these negative externalities, though it took time and hard work. In 1836, an English chemical manufacturer, William Gossage, constructed towers in which the gas was absorbed by falling water. The Leblanc process was finally replaced by carbonating towers that used ammonia to remove soda from salt by Ernest Solvay in 1861. The ammonia process was much more efficient than the Leblanc process not only in that it produced few undesirable by-products but also in that it was able to recycle the ammonia and carbon dioxide it used. By the mid 1860s the process was perfected and Solvay, like Perkin and Bessemer, became a very wealthy man.

Not all inventors in chemical engineering were that lucky. Charles Goodyear, the American tinkerer who invented the vulcanization process of rubber (1839) that made widespread industrial use of rubber possible, died deeply in debt. Another American, John Wesley Hyatt, succeeded in creating the first synthetic plastic in 1869, which he called celluloid. Its economic importance was modest at first due to its inflammability, and it was primarily used for combs, knife handles, piano keys, and baby rattles, but it was a harbinger of things to come.[55] The breakthrough in synthetic materials came only 1909, when the Belgian-born American inventor Leo Baekeland discovered bakelite.

Electricity. Like chemistry, electricity was a field in which *totally new* knowledge was applied to solve economic problems. From the beginning of the nineteenth century, the economic potential of electricity had been suspected. Humphrey Davy had demonstrated its lighting capabilities as early as 1808. Michael Faraday invented the electric motor in 1821 and the dynamo in 1831. Yet there was still substantial uncertainty about the possibilities of using electricity.[56] From the mid 1840s on, the earlier enthusiasm about electricity as a source of cheap energy waned. Its first effective use was not in power transmission, but in the telegraph. The telegraph was associated with a string of inventors, the most important of whom were the German S. T. von Soemmering, who demonstrated its capabilities in 1810; William Cooke, an Englishman who patented a five needle system to transmit messages (1837); and Samuel Morse who invented the code named after him that made the single-needle system feasible.

Electricity was, however, destined to become eventually a prime means of transmitting and using energy. To do this, several hurdles had to be overcome: an efficient way had to be devised to generate electric power using other sources of energy; devices to transform electricity back into kinetic power, light, or heat at the receiving end had to be created; and a way of transmitting current over large distances had to be developed. In addition, electricity came in two forms, alternating and direct current, and a decision had to be made

[55] Contrary to some accounts, Hyatt did not succeed in producing celluloid billiard balls (Friedel (1979), p. 52).

[56] When a visitor to the Royal Institute asked Faraday what use his newly invented toy had, he reputedly answered: what use is a newborn child? Thirty years later, in 1862, the seventy year old Faraday was taken on a tour of the English lighthouses and was able to see the beneficial effects of his "magnetic spark."

regarding which of the two forms was to dominate. For five decades inventors and engineers in Europe and the United States struggled in another international research effort, not orchestrated or coordinated, but guided into success by the ingenuity and perseverance of the participants.

Electric generators were crucial. Davy had shown in 1808 how electricity could drive an arc lamp, but apart from lighthouses it was not widely used in lighting. The principle of the self-excited generator was discovered in the mid 1860s by C. F. Varley and Werner von Siemens, and in 1870, the Belgian Z. T. Gramme had built a ring dynamo, which reduced the cost of alternating current substantially. Only then could the arc lamp be made practical, especially when in 1876, a Russian inventor, Paul N. Jablochkoff, invented an improved arc lamp (or "candle"), which used alternating current. Subsequently factories, streets, railway stations and similar places began to replace gaslight with arclight. In 1878, the American George Brush invented a high-tension direct-current lamp, which by the mid 1880s had come to dominate arclighting (Sharlin (1967)). The use of electricity expanded quickly in the 1870s: in 1879, at the Berlin exhibition, a miniature electric railway was displayed. In 1883 the industrial exhibition of Vienna demonstrated the use of electric blankets and hotplates, and electric streetcars were running in Frankfurt and Glasgow by 1884. The early 1880s saw the invention of the modern lightbulb by Joseph Swan in England and Thomas A. Edison in the United States. An electric motor using alternating current was built by Nikola Tesla in 1889, and together with George Westinghouse, the forces for alternating current defeated those advocating using direct current led by Edison. By 1890 the technical problems had been solved; electricity had been tamed.[57]

---

[57] Electricity found an unexpected use in electrochemistry, at this time largely confined to aluminum refining. The properties of aluminum had been known since the 1820s, and in 1855 the French chemist St. Clair Deville produced aluminum by electrolysis, selling it at about $200.00 per kg, which confined its use to jewelry and other esoteric applications. In 1886, a twenty-two year old undergraduate at Oberlin College, C. M. Hall, perfected a refinement method using molten cryolite and dynamos instead of Voltaic cells. Aluminum prices fell sharply. After K. J. Bayer invented a system to use bauxite and the age-hardening process was discovered by the German A. Wilm, aluminum became of major importance (Hall (1976); Multhauf (1967), pp. 477–79; C. S. Smith (1967c), pp. 599–600).

The development of steel, chemicals, and electricity required new scientific information for their perfection before they could become practical. Yet much technological progress in the years between 1830 and 1914 took the form of novel applications and refinements of existing knowledge. Moreover, the application of known techniques in new combinations often called for further inventions if the new idea was to become practical. As an example, consider the railroad. The railroad was not a proper "invention" because it was in essence nothing but a combination of high pressure engines with iron rails. The critical insight of combining the high pressure steam engine with the principle of the railway, which minimized the friction between wheel and surface originated from the genius of Richard Trevithick (1804). For two decades, engineers struggled with the unfamiliar problems of high pressure engines, the delicate balancing of heavy engines placed on iron rails, the driving-rod and crank mechanism connecting the piston to the wheels, the mechanics of suspension, the need to make stronger and more durable rails, and the design of efficient boilers. Once these problems were solved thanks to such engineers as Timothy Hackworth, George Stephenson, and his son Robert, the railway became inexorably one of the most potent forces of the nineteenth century. A variety of competing types of locomotives were designed, but the famous Rainhill competition in October 1829 determined without doubt that the Stephensons' *Rocket* had the superior design. It is clear, however, that the meteoric growth of railroads between 1825 and 1850 would not have occurred without subsequent refinements such as horizontal placement of the engine and improved axles, brakes, springs, couplings, the grading of roadbeds, tunnels, communications, and so on.

Similarly, the steamship contained few ideas that were truly new. The paddlewheel, which propelled the early steamboats, had been known in China for centuries. The first steamships were really sailing ships with auxiliary engines, with steam only helping out against unfavorable winds and tides. The *Savannah*, the first steamship to cross the Atlantic, used her engines for a total of 80 hours on a voyage of 30 days. It was not until fourteen years later, in 1833, that the *Royal William* made the voyage from Quebec to Gravesend largely by the force of steam power. The invention of the screw propeller by John Ericsson and Francis Smith in 1838 and the development of the compound expansion steam engine (first used on ships in 1854) were, however, important breakthroughs, making steamboats a serious, and eventually invincible, competitor to the sailing ship. In the second half of the nineteenth century the construction of ships shifted gradually from wood to iron. The British engineer Isambard K. Brunel designed the first large iron, screw-driven transatlantic steamships, the *Great Western* and the *Great Britain*. In the late 1870s, steel replaced iron as the main construction material. Here the transition was relatively quick: by 1891, over 80 percent of all ships under construction were made out of steel (Fletcher (1909), p. 281). Ships grew bigger, more powerful, and faster at unprecedented rates. While the typical ship, whether merchant or man-of-war, in 1815 was not much different from the typical ship of 1650, that of 1890 had little in common with its predecessor a half century earlier.

Nonetheless, it took decades for the new shipbuilding technology to supplant the old (Harley (1973)), in part because the production of wooden sailing ships was also experiencing rapid progress (Davis, Gallman, and Hutchins (1989)). Between 1820 and 1860, sailing vessels were completely redesigned. Rigging and sails were revamped, resulting in ships that were faster yet required fewer sailors. Ropes were supplanted by more durable metal cables, and wooden fittings by metal fittings. Ships became more elongated and streamlined, and sails became smaller and more numerous, culminating in the famous tea clippers built by American and British shipbuilders after 1850, which could attain amazing speeds. The fastest clippers could sail from China to Britain in three months (Chatterton (1910), p. 270).

On some occasions, a technical solution looked simple, and it may seem surprising at first sight that it took so long until producers got it right. During the entire nineteenth century, mechanics experimented with a device that would allow an individual to propel himself rapidly while seated. Yet it was not until John K. Starley, a Coventry mechanic, built the Rover safety cycle in 1885 that the balanced position and easy steering of today's bicycles became feasible. The flat-link-chain drive of these bicycles contained nothing that would have been beyond the ability of, say, a Henry

Maudslay or a Richard Roberts.[58] The bicycle, in some views, prepared the way for the automobile and the motorcycle, and supplied a form of cheap and democratic yet personal transportation. Hiram Maxim, the inventor of the Maxim machine gun (as well as an unsuccessful steam-driven aeroplane) pointed out that "the bicycle created a new demand that was beyond the ability of the railroad to supply. Then it came about that the bicycle could not satisfy the demand which it had created . . . the automobile was the answer" (cited by Cardwell (1972), p. 199–200).

The classical case of a novel combination of known techniques, laced with a number of original contributions, was the development of the automobile. The internal combustion engine was first suggested by Christiaan Huygens in the seventeenth century. In 1824, Sadi Carnot had described the limitations of the steam engine as an energy source and pointed to heated air as the best potential means to generate motive power. Prolonged research efforts notwithstanding, it turned out to be infeasible to use steam power for carriages and agriculture (Evans (1981)). During the nineteenth century dozens of inventors, realizing the advantages of an internal combustion engine over steam, tried their hand at the problem. A working model was first constructed by the Belgian Jean-Etienne Lenoir in 1859 and perfected in 1876, when Nicolaus August Otto built a gas engine using the four-stroke principle. Otto, originally a traveling salesman, worked on the problem from 1860 on, after he read in a newspaper about Lenoir's machine. He was an inspired amateur, without formal technical training. The problem was solved eventually with the four-stroke engine, but initially he saw this as a makeshift solution to the problem of achieving a high enough compression. It was not until later that Otto's four-stroke engine, which is still the heart of the automobile engine, was acclaimed as a brilliant breakthrough (Bryant (1967), pp. 650–57). Interestingly, the four-stroke engine had been recognized earlier as the only way in which a Lenoir-type engine could work efficiently. Using Carnot's

[58] The pneumatic tire, on which the comfortable bicycle ride depends, had been patented in 1845 and tested successfully. Yet the needs of horse carriages were apparently insufficient for its widespread production, and the invention was soon forgotten until a Belfast veterinary surgeon, J. B. Dunlop, who was unhappy with the comfort of his ten-year old son's tricycle ride, resurrected it in 1888.

thermodynamic theory, Alphonse Beau de Rochas specified in 1862 an Otto-type principle to describe an efficient internal combustion and obtained a patent for his idea, although it was never actually put into effect. Otto's invention, however, was quite unaffected by Beau de Rochas' theoretical insight.[59] The "Silent Otto," as it became known (to distinguish it from a noisier and less successful earlier version), was a huge financial success. The advantage of the gas engine was not its silence, but the ability of the operators to turn it on and off at short notice, whereas steam engines took a long time to develop pressure.

Otto's invention was in the mold of the classical invention of the earlier Industrial Revolution years and his career parallels that of James Watt in some important aspects. Both introduced a crucial improvement into an existing machine, which turned it from a comparative rarity into a dominant technique. Both were successful in large part due to the trust and business acumen of their partners, Matthew Boulton in the case of Watt, Eugen Langen in the case of Otto. Both became wealthy thanks to patent protection, although one of Otto's patents was invalidated in 1886 on the ground that Beau de Rochas had preceded him. Both had contact with scientists, but were eventually inspired in their crucial breakthrough by sheer intuition.

Otto's gas engine was soon to adopt a new fuel. Somewhat earlier, in the 1860s, the process of crude oil refining using a method called cracking was developed. At that time the main interest was in lubricants, paraffins, and heavy oils, with "petrol" or gasoline considered a dangerously inflammable by-product. In 1885 two Germans, Gottlieb Daimler (who had previously worked for Otto) and Karl Benz, succeeded in building an Otto-type, four-stroke gasoline-burning engine, employing a primitive surface carburetor to mix the fuel with air. Benz's engine used an electrical induction coil powered by an accumulator, foreshadowing the modern spark-plug. The Dunlop pneumatic tire, first made for

[59] Otto apparently never read Beau de Rochas' pamphlet, and at first misunderstood the principles that made his invention work. As Bryant (1966) remarks, his invention was another example of how good engineering could be based on bad science.

bicylces, soon found application to the automobile. In 1893 Wilhelm Maybach, one of Daimler's employees, invented the modern float-feed carburetor. Other technical improvements added before 1900 included the radiator, the differential, the crank-starter, and pedal-brake control. The dazzling development of the automobile in a few years is testimony to the technological creativity of the Western world during the second Industrial Revolution. The automobile and the bicycle had an effect on technology similar to the mechanical clock five centuries earlier: mechanics involved in making and repairing them acquired the skills and the ideas to extend the principles involved to other areas.[60]

In steam technology, improvements in fuel efficiency and design continued, although progress encountered diminishing returns and fewer revolutionary advances were made. The American George Corliss made many improvements to the steam engine, culminating in the famous Centennial engine shown at the Philadelphia Centennial Exhibition in 1876, the largest and most powerful every produced. More revolutionary in concept was the steam turbine, invented by Charles Parsons in 1884, which returned to the idea of Hero's aeolipile. The problem Parsons solved, contemporaneous with the Swede Gustav de Laval, was that reciprocating steam engines could develop only limited speed.[61] For electrical generation and ship propulsion much higher speeds were needed, and the turbine provided them: the prototype that Parsons built in 1884 ran at 18,000 rpm and had to be geared down. By 1900, the turbine had become a serious rival to regular steam engines. Parsons' engineering genius was heavily indebted to his scientific background, and his example should warn us against quick generalizations about the decline of British technology in this period due to scientific

[60] As Rosenberg (1976, p. 24) has suggested, much of the experience accumulated in bicycle production was put to good use in the automobile industry. The best known case in which bicylce makers went on to bigger things is of course that of the Wright brothers, builders of the first operational airplanes.

[61] The turbines built by Parsons and de Laval were quite different in design. The history of the turbine serves as a good warning against easy generalizations about inventions emerging "when the time was ripe." Although Parsons and de Laval arrived at their inventions almost simultaneously, no fewer than two hundred patents for gas and steam turbines had been taken out in Britain alone during the century before 1884. None of these inventions, apparently, were satisfactory.

backwardness. If science owed much to the steam engine in the first half of the nineteenth century, it repaid its debt with interest in the second half (Hall (1978), p. 99).

By the late nineteenth century, the steam engine was coming under attack from another corner: Rudolf Diesel built an internal combustion engine based on the idea that the temperature of air inside a combustion chamber could be raised sufficiently by compression to ignite the fuel, thus converting the energy of combustion entirely into work. In 1897 he built the first engine that burned heavy liquid fuel, and after a decade of further development and improvements, Diesel engines began to challenge steam everywhere. The Diesel invention is paradigmatic of the Second Industrial Revolution. Like James Watt and Richard Trevithick, Diesel was a "rational" inventor, in search of efficiency above all else. He was not a tinkerer, however, but a trained engineer, working with state of the art scientific techniques. He started off searching for an engine incorporating the theoretical Carnot cycle, in which maximum efficiency could be obtained by isothermal expansion so that no energy is wasted, and a cheap, crude fuel can be used to boot (originally Diesel used coal dust in his engines). Isothermal expansion turned out to be impossible, and the central feature of Diesel engines today has remained compression-induced combustion, which Diesel had considered at first to be incidental (Bryant (1969)).

From a purely economic point of view, it could be argued that the most important invention was not another chemical dye, better engines, or even electricity, since most of the inventions described—with the exception of steel—had serviceable substitutes, though not as efficient or as cheap. There is one innovation, however, for which "social savings" calculations from the vantage point of the twentieth century are certain to yield large gains. The so-called "American" system of manufacturing assembled complex products from mass-produced individual components. Modern manufacturing would be unthinkable without interchangeable parts. The term "American" is somewhat misleading: the idea that interchangeability had enormous advantages in production and maintenance had occurred to Europeans in the eighteenth century, and French and later British engineers fully realized its economic possibilities.[62] Moreover, what was regarded in the 1850s as "the American System" was not exactly interchangeability, but the application of high-quality, specialized machine tools to a sequence of operations, particularly in woodworking, as well as higher operating speeds and sequential movements of materials. As Ferguson (1981) has pointed out, mechanized mass production and interchangeable parts were not identical, and the former did not imply the latter.

Interchangeable parts was not a proper "invention." It was to become eventually a vastly superior mode of producing goods and services facilitated by the work of previous inventors, especially the makers of accurate machine tools and cheap steel. To be truly interchangeable, the parts had to be identical, requiring high levels of accuracy and quality control in their manufacture. It is now realized that full interchangeability was more difficult to achieve than had previously been believed. The use of interchangeable parts grew slowly after 1850, and recent research has shown that the American System was adopted far more haltingly and hesitantly than has hitherto been thought. Many American firms, such as McCormick, Singer, and Colt, owed their success to other factors and not to complete interchangeability (Hounshell (1984); Howard (1978)). At first, goods made with interchangeable parts were more expensive and adopted mostly by government armories, who considered quality more important than price (Howard (1978); M. R. Smith (1977)). Only after the Civil War did American manufacturing gradually adopt mass production methods and Europe soon

[62] In point of fact, one could argue that moveable type was the first application of interchangeable parts. As early as 1720, the pioneering Swedish engineer Christopher Polhem used interchangeable parts. In 1785 Thomas Jefferson, then ambassador in Paris, reported that a French gunsmith named Honoré Blanc was making musket locks entirely from interchangeable parts. Like many other French technological advances, Blanc's was buried by jealous competitors: pressure from gunsmiths forced the closure of his arsenal in 1797. For Britain, too, the textbook tale that Europe lagged behind the United States in the adoption of interchangeable parts is largely a myth. In 1832, the London mechanic Bryan Donkin built a continuous-web paper machine that astounded American visitors over the ease at which repairs could be made by replacing a defective part with another taken from the shelves. Joseph Whitworth was a tireless advocate of interchangeability from the 1840s (Musson (1975a)). Other British engineers who experimented with standardized parts were John Bodmer, James Nasmyth, and Daniel Gooch, a manufacturer of locomotives (Musson (1981)).

followed. First in firearms, then clocks, pumps, locks, mechanical reapers, typewriters, sewing machines and eventually engines and bicycles, interchangeable part technology eventually proved superior. Although in the long run true interchangeability was inexorable, its diffusion in Europe was slowed down by two factors: its inability to produce distinctive high quality goods, which kept consumers long faithful to skilled artisans, and the resistance of labor, which realized that mass production would make its skills obsolete (Cooper (1984)).

Of related importance was the development of continuous-flow production, in which the worker remained stationary while the tasks were moved to him. In this way, the employer could control the speed at which operations were performed and minimize the time wasted by workers between operations. The American Oliver Evans is generally credited with this invention at his justly famous grainmill in rural Delaware. A combination of Archimedes screws and endless belts moved the grain from task to task in fully automated fashion. Milling grain, or for that matter slaughtering cattle, as Siegfried Giedion has observed, did not involve assembling but disassembling a product. Most manufacturing entailed assembling and combining, and the continuous-flow process was eventually adopted there as well using endless belts which moved the materials and parts among workers. The first documented occurrence of assembly-line production was in a biscuit factory in Deptford, Britain in 1804. In 1839, J. G. Bodmer's machine tool workshop was organized on similar principles. Yet it was not until the last third of the nineteenth century that continuous flow processes were adopted on a large scale, especially in the great stockyards of Chicago and Cincinnati (Giedion (1948)).

The standard of living of the masses depended, above all, on nutrition, and here advances in agriculture and food processing were of primary importance. In this context the term "free lunch" may be interpreted almost literally. The new technologies of the nineteenth century affected food supplies through production, distribution, preservation, and eventually preparation.

In agriculture, the adoption of the new husbandry based on fodder crops and stall-fed livestock continued apace, though in France and in most of Eastern Europe progress was slow. New implements and tools appeared on the scene, but here the traditional obstacles to agricultural progress slowed down growth: inventions that were useful in some environments failed elsewhere. The mechanical reaper, invented by the American Cyrus McCormick in 1832 for the flat, wide open midwestern fields, had to be modified extensively before it was suitable to Europe (David (1975), pp. 233–75). Of the new crops adopted after 1830 or so, the sugarbeet deserves mention, after a French engineer, Benjamin Delesert, discovered the beet sugar refinement process (1812). Only after 1840 did beet sugar become a serious threat to cane sugar.

Agricultural productivity owed much to the extended use of fertilizers. Farmers learned to use nitrates, potassium, and phosphates produced by the chemical industries. In addition, guano (bird droppings) were imported in large quantities from Peru. In the United States, the large stockyards produced fertilizers made from animal bones combined with sulphuric acid. The productivity gains in European agriculture are unthinkable without the gradual switch from natural fertilizer, produced mostly *in loco* by farm animals, to commercially produced chemical fertilizers. Fertilizers were not the only scientific success in farming: the use of fungicides, such as Bordeaux mixture, invented by the Frenchman M. Millardet in 1885, helped conquer the dreaded potato blight that had devastated Ireland forty years earlier. The Industrial Revolution affected agriculture in many other ways. Steel implements, drainage- and irrigation pipes, steam-operated threshers, seed drills, and mechanical reapers slowly but certainly improved productivity and expanded the supply of food and raw materials. Yet here, more than anywhere else, the old resisted the new, and modern tools and techniques continued to coexist with manual operations that had not changed in centuries. Mechanizing agriculture involved some major technical difficulties. Much work in agriculture, such as weeding, picking, and milking, was carried out by movements of the human fingers, as opposed to the sweeping or beating motions of the human arm. We have already seen how hard it was to substitute technology for human fingers in the spinning of yarn; much of agriculture was subject to the same kind of difficulties.

The other reason for the slowness of the mechanization of agriculture (compared to manufacturing) was the lack of power substitutes. In most industrial processes, the act of manufacturing can occur at the site of the power source. The utilization of more efficient energy sources was thus rather simple. In agriculture, the power sources had to be brought to the production site (i.e., the land) for most activities, and thus plowing, harrowing, reaping, raking, and binding remained dependent on draft animals long after manufacturing and transportation had adopted the steam engine.[63] Only when the work could be carried out near the power source was mechanization relatively early: the threshing machine built by the Scotchman Andrew Meikle in 1784 spread quickly, as did the winnowing machine built by James Sharp in 1777. These machines were attached to steam engines in the first half of the nineteenth century, but they remained something of an exception. The internal combustion engine solved all that, and by the eve of World War I, tractors and combines were being introduced on both sides of the Atlantic. In 1880, it still took an average of 20 man-hours to harvest an acre of wheat in the United States; by 1935 this figure had fallen to 6.1.

Of special interest to the historian interested in economic welfare is the development of food preparation and preservation. Immeasurable human suffering has been caused over the ages by nutritional deficiencies and by people unwittingly consuming contaminated foods. The idea of preserving food by cooking followed by vacuum-sealing was hit upon by Nicolas Appert in 1795. Appert originally used glassware to store preserved foods, but in 1812 an Englishman named Peter Durand suggested using tin-plated cans, which were soon found to be superior. By 1814, an engineer named Bryan Donkin was supplying canned soups and meats to the Royal Navy. Canning, too, was a good example of technology leading science, because only after Louis Pasteur's pathbreaking discoveries was it understood why canning worked, and not until the end of the century did it become clear that the optimal cooking temperature was about 240°F. Canned food played an important role in provisioning the armies in the United States Civil War, and led to vastly increased consumption of vegetables, fruit, and meat in the rapidly growing cities.

[63] Attempts to use steam power on the fields produced both John Fowler's cable plowing (1850) in Britain and steam tractors in the United States. Yet steam power in agriculture, except for stationary threshing machines, was not a lasting success.

Other food preservatives were also coming into use. Gail Borden invented milk powder in the 1850s and helped win the war for the Union and himself to a fortune. By the end of the century his dehydration idea was also successfully applied to eggs and soups. The centrifugal cream separator, invented by Gustav de Laval in 1877 soon became the cornerstone of the dairy cooperatives in Denmark, Holland, and Ireland. Pasteur himself showed how to sterilize bottled cowsmilk, making one of nature's most wholesome foods accessible to millions of protein-starved urban children. Cooling was an alternative form of preservation. In the eighteenth century, ice was preserved in special icehouses, and an international market in ice emerged in the early nineteenth century, though in the warm seasons the price of ice was so high that only the rich could afford it. Mechanical refrigeration was gradually developed and improved upon between 1834 (when the first patent for the manufacture of ice was issued in Britain) and 1861 (when the first frozen beef plant was set up in Sydney, Australia). By 1870, beef transported from the United States to England was preserved by chilling (29°–30°F). The efficient method of preserving beef, however, was by deep freezing at about 14°F. In 1876 the French engineer Charles Tellier built the first refrigerated ship, the *Frigorifique,* which sailed from Buenos Aires to France with a load of frozen beef. By the 1880s, beef, mutton, and lamb from South America and Australia were supplying European dinner tables. Farming in Europe suffered from this competition, but the consumer, the ultimate and final arbiter of all questions of economic progress, benefitted greatly.

Two industries that played a central role in the early stages of the Industrial Revolution experienced comparatively slow development after 1830, yet continued to score solid productivity gains. In mining, a major technical difficulty was overcome when the advantages of compressed air were recognized. Between 1849 and 1856 pneumatic mining tools such as circular cutters for underground hewing were devised. Yet not before the end of the century did power tools definitively supersede the pick and the shovel.

In textiles, one major innovation stands out: the sewing machine. Apparel making had lagged behind the rest of the textile industry during the early stages of the Industrial Revolution. Elias Howe, an American, is usually credited with its invention, but he merely added one crucial feature, the lock stitch. The important insight embodied in the sewing machine was that the machine should not mimic the human hand but use a double thread, with the eye of the needle far down.[64] The man who deserves the most credit for perfecting the sewing machine is Isaac Merritt Singer, who powered his machine with a foot treadle. It was later adapted to make shoes (1861) and carpets (1880). Unlike many other inventions in textiles, the sewing machine did not lead to factories, as it did not require a centralized power source. Instead, it kept struggling domestic workers (mostly women) occupied, and created a system of notorious "sweatshops." The sewing machine was also unusual because it was slow in adopting interchangeable parts; the Singer company was successful primarily because of brilliant marketing, but still used skilled fitters long after its competitors had switched to fully interchangeable parts (Hounshell (1984), pp. 67–123).

In the rest of the textile industry, the period after 1850 completed what had been begun earlier. The combing of wool, which had long defied mechanization, was further improved by the introduction of the Lister-Donisthorpe nip machine in the 1850s, reviving the fortunes of the Yorkshire worsted industries. The Heilmann combing machine was used widely on the Continent, especially in Alsace. In spinning, the throstle that had dominated the scene until the 1860s was slowly superseded by ring spinning, invented by a Rhode Island mechanic, John Thorp, in 1828. Ring spinning differed from mule spinning. In the latter, the twist was imparted by the combined action of rollers and the revolving spindles. In ring spinning, the twisting of the yarn was carried out by the circular movement of a clip on the rapidly turning ring, called the "traveler." The traveler guided the thread and assured its winding on the spindle. Ring spinning was continuous rather than intermittent, and required less skill and strength than mule spinning. It produced, however, a slightly inferior product, especially for fine

[64] The inventor with the best claim to the sewing machine was the Frenchman Barthélemy Thimonnier, who devised a working model in 1830 using a chainstitch system that made it a combination sewing- and embroidery machine. Thimonnier ran into violent opposition from tailors. His army uniform factory in Paris, in which he had the new devices installed, was raided twice in the 1840s, and the machines were destroyed. Like many other French inventors, he died a poor and bitter man.

yarns. Ring spinning did not spread widely until the second half of the nineteenth century, but then it conquered the American textile industry rapidly. Lancashire in Britain remained loyal to mules, however, a fact that has long intrigued economic historians.[65] In weaving, the power loom continued to replace hand loom weavers after 1850, but automation came to weaving only when J. H. Northrop built the first automatic loom in 1894. Within twenty years it did to the United States weaving industry what Roberts' self-actor had done to British spinning in 1830. By this time, however, Britain's textile industries had lost their position at the cutting edge of technology, and their adoption of the Northrop loom was slow.

In the textile industries, mechanization during the second half of the nineteenth century had the character of mopping-up operations. Industries that had been skipped for one reason or another by the Industrial Revolution were now caught up in its maelstrom. A good example is the hosiery industry. Lee's knitting frame, invented in 1589, remained at the heart of the stocking-frame technology until deep into the nineteenth century. The only major improvement introduced before was an attachment for knitting ribbed fabrics, invented by Jedediah Strutt, who later became Arkwright's partner in 1758. It is often maintained that technical progress in frame-knitting was impeded by low wages, as workers had essentially no alternative employment (Wells (1958), p. 598). Another alleged cause for the lateness of mechanization was the riotousness of the Nottingham knitters, which may have scared off potential innovators. Such factors, however, were also present in other industries such as wool where they seem to have had little effect in impeding progress. The main problem in applying power to the knitting industry was purely technical. It was complicated to adapt the fine mechanism which varied the tube width, on which the quality of the stockings depended, to more powerful machinery. When the technical problems were finally solved by Matthew Townsend in the late 1840s and William Cotton in the early 1860s, hosiery mechanized just as fast as any other industry.

Finally, I turn briefly to what may best be called information

[65] See Sandberg (1974) and Lazonick (1981) for different interpretations of the diffusion of ring spinning.

processing technology, though the term rings vaguely anachronistic. Following the telegraph in the 1830s, a "modern" pattern starts to emerge, in which practice followed theory. Thus Hermann Helmholz, a German physicist, experimented with the reproduction of sound, which inspired a Scottish-born speech therapist named Alexander Graham Bell in his pathbreaking work on the telephone (1876). Supplementary inventions such as the switchboard (1878) and the loading coil (1899) made this one of the most successful inventions of all time. Similarly, wireless telegraphy was implicit in the theories of James Clerk Maxwell, from which Heinrich Herz deduced a more general theory of waves. Guglielmo Marchese Marconi combined the theories of these ivory tower theorists into wireless telegraphy (1895), and Lee DeForest and R. A. Fessenden showed how wireless radio could transmit not only Morse signals but sound waves as well (1906).

Little science was necessary for another invention that had a huge effect on information technology: the typewriter. The idea of the typewriter is conceptually obvious (the first prototypes were aimed at the use of the blind), but a number of minor technical bugs bedeviled its perfection (such as bars clashing when two letters were typed very closely together). These problems were finally solved by C. S. Sholes, who sold his patent to the Remington Co. in 1874, and a small revolution in the office began. The technical problems in the printing industry were more complex. Typesetting had always been laborious and slow work, and the need for improvement was emerging all the more acutely with the rise in literacy and the greater thirst for information in the late eighteenth and nineteenth centuries. The use of cylinders in printing was first pioneered in cotton-printing around 1780. Cylindrical impression and inking was applied successfully to newspaper and book printing by a German immigrant in London, Friedrich Koenig, in 1812. The first rotary press was built in Philadelphia in 1846. A horizontal cylinder contained the printed material, and rotated in contact with smaller cylinders, each of which corresponded to a page, with automatic grippers guiding the pages from cylinder to cylinder. This machine, originally conceived by Robert Hoe, found its way to Europe where many leading newspapers adopted it. It was fast but labor intensive. Typecasting, equally revolutionary in that it recast the type each time anew using an automatic process, was perfected in the United States in 1838, and by 1851 it had spread all over Europe. An alternative technique, invented by Henry Bessemer (of steel fame), was the "pianotype," where an operator worked at a keyboard. For a while, a confusing multitude of automatic typesetting techniques were in operation at the same time. Between 1886 and 1890, a German immigrant to the United States, Ottmar Mergenthaler, invented the linotype machine, which cast and set a whole line at a time using a keyboard controlling hundreds of "matrices" from which the letter molds were made. Linotype machines were primarily used for newspapers; for books, a related machine, the monotype typesetter, was developed in the 1890s.

Pictures as well as written words underwent changes. In 1798 a Polish typographer by the name of Alois Senefelder invented lithography to reproduce drawings. Photography developed gradually in the nineteenth century, exploiting the well-known property of silver salts to blacken under the influence of light. Two Frenchmen, Joseph N. Niépce and Louis Daguerre, discovered in the 1830s how to make the pictures permanent, and the Englishman W. H. Fox Talbot succeeded in taking photographs on paper (patented in 1841), so that many copies could be printed. Eventually George Eastman introduced his Kodak in 1888, which helped make photography one of the most durable and widely-enjoyed hobbies. In the technology of papermaking, machinery producing continuous sheets had been patented by the Frenchman Louis Robert in 1798. Robert's machine was further developed by Bryan Donkin and eventually came to be known as the Foudrinier machine, after a London Stationer (Clapham (1957), p. 416). With the increase in demand for paper, new raw materials became necessary and, after much experimentation, the use of woodpulp was perfected in about 1873.

* * *

Technology in the twentieth century has developed so rapidly and become so complex that no justice can be done to it here. The trend toward a more "scientific" approach to technology continued, and many steps forward would not have been possible without advances in mathematics, physics, chemistry and biology that occurred after 1870. Three hundred years after Francis Bacon, his dream of advancing technology and material welfare by constantly increasing their scientific basis had become reality. Yet the nonscientific taproots of invention—serendipity, luck, and inspiration—did not disappear, and probably never will. The lonely individual inventor will not be superseded entirely by the corporate research team, any more than the brilliant general will be replaced by a war-game-playing computer. Technological progress is more efficient today in the sense that fewer false turns are taken, and blind alleys are easier to avoid. But the twentieth century still runs into two dilemmas common to all ages. The first was rooted in the fact that some devices can be made to work long before it is understood why or how they work. Especially in medicine and biological technology, such empiricist technology ("try every bottle on the shelf") is still rampant, and it explains for instance our painfully slow advances in cancer therapy. The second may be called the Da Vinci problem: gadgets and devices can be conceived that are known to be possible, but cannot be built efficiently because supporting technologies are lacking. Energy generation without long-term environmental damage, safe and effective contraception, insect control, and superfast long-distance travel are among such technological bottlenecks of our own age.

The difference between the nineteenth and the twentieth centuries is the technological optimism of our age. In spite of immense progress, the nineteenth century was an age in which technology was believed to be constrained and ultimately incapable of lifting mankind out of its poverty. The sources of pessimism were varied. Most economists believed with Ricardo that the standard of living would eventually be set by the minimum of subsistence, or some other ceiling governed by demographic factors. As late as 1890, Alfred Marshall's *Principles* epitomized economics as the study of small, continuous changes rather than abrupt breakthroughs. Despite a century and a half of innovations, many of which were abrupt and revolutionary enough, Marshall still believed with Leibnitz that Nature does not make leaps. Economists were not alone in this. Physicists had discovered the Laws of Thermodynamics, and with them the limitations on energy generation. Geologists began to draw attention to the limits of the earth's resources leading some economists like W. S. Jevons into fallacious predictions of an imminent energy crisis. In a related context, the theory of evolution suggested that long-term change was typically slow and gradual. The twentieth century, on the other hand, has understood that technology is limitless and can advance in leaps and bounds, and that only society's proclivity to destroy the conditions for its growth halts progress. Twenty years after Marshall, Schumpeter wrote about "spontaneous and discontinuous changes," while a new physics, in which critical masses and quantum leaps occupied a central position, was emerging.

By 1914 the Western world had grown for many decades, and the gap between it and the rest of the world had become huge. The sources of economic growth had been the triad of gains from trade, capital accumulation, and technology. The gains from trade on all levels were still rising, but as transportation and information networks became more efficient, the marginal effects of trade fell. Moreover, Smithian growth is always vulnerable to political events. In the late 1870s, growing protectionism partially offset the gains that the railroad and the steamship brought; in August 1914 it all fell apart. Capital widening raised incomes, but by 1900 capital from the technologically progressive countries (Britain and France) was flowing out to newly settled areas in North America and Australia, or to comparatively less developed European economies, suggesting either strongly diminishing returns, distorted and inefficient capital markets, or both. In any case, capital accumulation on its own was an equally weak reed for long-term economic change. Technological progress alone can support sustained growth, because it alone has not run into diminishing returns. And yet technology, in its own way, was a mixed blessing. World War I showed the dangers of the second Industrial Revolution. It was a war of steel, chemicals, and internal combustion engines. The inventions that had the power to bring unlimited prosperity could also bring unlimited misery.

Technological change, then, accounted for *sustained* growth. It was not caused by economic growth, it caused it. It had no substitutes. Economic historians have been fond of judging particular innovations in terms of their forward linkages, that is, the savings in cost in using a new technique compared to the next best alternative. Some innovations, such as the railroad or the steamship or ring spinning, are put in their rightful modest place this way. The printing press, Cort's rolling and puddling process, and chemical fertilizers probably look more important. But taken ensemble, the plethora of innovations occurring between 1750 and 1914, whether narrowly technical or organizational, obey the "indispensability axiom." Had there been no technical change, other forms of economic progress would eventually have ground to a halt, and Europe would have ended up like the Roman or Chinese Empires. The question is: why didn't it?

7. THE HISTORICAL ROOTS OF TECHNOLOGICAL CREATIVITY

What I have surveyed above is a tale of technological success. It may seem to some "Whiggish", telling a story of progress and improvement. Although technology did not move linearly, and not even necessarily always unidirectionally, it can scarcely be maintained that today's techniques for producing *any* good or service, from wheat to filling a tooth, are not vastly superior to those of 1750 let alone the year 1000. "Better" here is meant strictly in the tautological economic sense of being more efficient. Much, though by no means all, of the progress occurred in Europe. The survey above has therefore been inevitably Europocentric. The notion, however, that modern technology and the modern economy it created were primarily the responsibility of occidental Europe is oversimplified. Thanks to the labors of Joseph Needham and his associates, we know now that China had developed by about 1400 a highly sophisticated economy. Needham has tirelessly tried to link almost every major invention made in the West to Chinese precedent, and although in some of these cases we have to assume that Europeans made some inventions independently, the priority of Chinese technology in many areas can hardly be disputed. Europe also owed debts to other non-European civilizations, from India to pre-Colombian America. In the last century it has repaid these debts with usurious interest, and its phenomenal success is expressed in recent titles such as *How The West Grew Rich* and *The European Miracle.*

It is somewhat misleading to look at Europe and its offshoots as a single unit from which technological advances emanated. At all times, only a small part of the West was truly innovative. The center of gravity kept changing over the centuries: from Italy to southern Germany to the Low Countries, from there to Britain and France, and in the late nineteenth century to Germany and the United States. If anything is to be learned from our historical survey, it is that technological creativity was not only rare but ephemeral. Most societies that ever existed were not technologically creative, and those that were, were so for relatively short periods. However, even those European economies which were not innovative were often creative in the sense that they adopted new techniques developed elsewhere. Whether the novel techniques were developed in Britain, France, or the Low Countries, they sooner or later penetrated Scandinavia, Russia, and Southern Europe. The factors behind this unique technological creativity are complex, and cannot be done justice to here.[66] Instead, I will try to take up Heertje's challenge cited in the introduction to this survey, and ask how we can come to grips with technological progress as an economic phenomenon dominating economic history in the past twenty-five centuries.

The framework and historical survey outlined above suggest that technological change is sufficiently heterogeneous to defy simple generalizations. That does not imply that I agree with the undertone of defeatism implicit in Heertje's statement, that technological change cannot easily be explained. What is important is to separate the different forms technological change takes, and to emphasize that the different forms are subject to different forces. A theory of technological change should be as simple as possible but, as Einstein said in a different context, not simpler. The first distinction we have to make is well known: there is a difference between being imitative and innovative. An imitative society will adopt new techniques wherever they come from, as long as the technique is demonstrably superior. An innovative society is capable of creating new techniques, but does not necessarily apply them itself with great success. Innovation and imitation were both manifestations of technological creativity, but they did not always coincide precisely. Until recently, for example, it was widely believed that the industrial nations of East Asia were only capable of technological creativity through imitation, but could not innovate anything themselves. We know better now, but the prejudice is symptomatic of the difference between the two. Similarly, during the late eighteenth century, France and Britain were both innovative in the sense

[66] See Mokyr (1989b) for a more complete analysis of the factors determining technological creativity.

that they generated a stream of inventions amounting to what we call today an industrial revolution. But although both countries were innovative, Britain was by far more creative. Inventions were adopted faster, and the old economic status quo was overthrown at a quicker pace. In its early decades, the United States was imitative rather than innovative (Wilkinson, 1963). Another society which appears to be more imitative than innovative was early medieval Europe. It adopted, improved, and refined technologies it inherited from the Romans or borrowed from the Middle East or the Orient. Until it invented the weight-driven clock and the spinning wheel, it made few major inventions of its own.

Technologically creative societies are not necessarily innovative, but experience shows that imitation usually tends to lead to innovation. David Landes has pointed out that "good innovators make good imitators." The reverse is also true: many technologically creative societies started as imitators, and eventually learned to generate genuine innovations until their former teachers had to humble themselves and emulate the emulators. There is adequate historical precedent for the Nova plant in Fremont, CA where General Motors is assembling its best car, a Toyota clone. Even Britain during the Industrial Revolution, one of the most technologically creative societies ever, was as much a follower as a leader, and some of its most radical innovations were French and German imports. After 1815, however, British inventors from Richard Roberts to Percy Gilchrist established Britain firmly as the world's most innovative economy—for a while.

Yet inventions differ qualitatively from each other. I have earlier established the difference between microinventions and macroinventions. The former are basically improvements *within* a given technique. They consist of mostly minor improvements, adaptations to local circumstances, and debugging of devices based on the right idea but not quite workable for a variety of small causes. Macroinventions, on the other hand, create new techniques, the technological equivalent of what evolutionary biologists call speciation. They constitute a clear-cut discontinuity with the past, and are, in the evocative metaphor of the biologist Richard Goldschmidt, "hopeful monstrosities." Indeed, some of the most radical inventions like Newcomen's atmospheric engine, Montgolfier's hot-air balloon, and Bessemer's converter, must have appeared exactly thus:

hopeful to the inventor, but monstrosities to the unaccustomed eyes and ears of his contemporaries.[67]

Technological change occurs on a number of levels. On one level it continually improves and refines *existing* techniques, by a process akin to anagenesis (gradual change within species). On a higher level, techniques themselves are born and die. For instance, it was hand loom weaving as a technique, and not the hand loom weavers as individuals, that practically disappeared in the nineteenth century. One of the most important insights of the Gould–Eldredge school of evolutionary biology is to recognize that evolution occurs in hierarchies (Gould, 1982). Species change gradually in a continuous accumulation of micromutations. At a higher level, however, macromutations and extinction create and destroy entire species. I propose precisely such a hierarchy in technological progress to sharpen our understanding of the economic roots of technological creativity.

It is possible to classify inventions into the two categories? As always, there is a gray area. But in the majority of cases, I think that one's verdict—though inevitably based on intuition and judgment more than on any hard rules—is fairly unambiguous. Macroinventions rarely emerged in one single blow (though it did happen, as in the case of Gutenberg's moving type or Perkin's aniline dye). More often they appeared in stages, yet were discontinuous enough not to be confused with microinventions. Consider the steam engine: between Papin and Trevithick a number of clear-cut steps can be identified, and so the question "who invented the steam engine" has no single answer. Yet the emergence of the steam engine created a "species" which was not in existence a century before. No tale of minuscule cumulative improvements can fit these facts alone, though microinventions inevitably followed macroinventions. Compare this with the tale of the invention of McCormick's reaper: though it was invented in the early 1830s, it did not come into general use until a series of small microinventions rendered it workable (Olmstead, 1975). A similar story can be told

[67] The classification used here differs from that proposed by Usher (1955). Usher defined a primary invention as one that was not carried to a stage of general commercial use, whereas secondary and tertiary inventions applied it to practical and commercial use.

about the power loom, the steamship, and the bicycle. All these macroinventions required subsequent microinventions to iron out the wrinkles and become viable. It seems obvious to classify the combination of the Hargreaves–Arkwright–Crompton mule as a macroinvention, as it produced a technique quite dissimilar to what was there before. But does Richard Roberts' self-actor qualify as a macroinvention? Not all microinventions were necessarily minor and nameless. As long as they altered the existing technique but did not create a new one, they cannot be termed macroinventions. Thus the self-actor does not qualify, nor does the compound steam engine. When is an invention just an improvement on the old, and when is it a new technique? Unfortunately, we have no good taxonomical system in technological history. Judgment and common sense will have to remain our guide here.[68] In Mokyr (1989a) I offer five examples of what I consider to be examples of quintessential macroinventions made during the Industrial Revolution. They are gaslighting, chlorine bleaching, the breastwheel, the Jacquard loom, and hot air ballooning. All five gave rise to a technique that was in numerous crucial aspects radically different from everything that came before.[69]

It is further useful to distinguish two kinds of macroinventions. One is what may be called a radical macroinvention, in which literally a new idea is conceived and implemented. This is the closest analogue to a mutation in biology, in which through some genetic accident a new species emerges. Of the examples mentioned

[68] It is, perhaps, some consolation that similar ambiguities exist in evolutionary biology. Evolutionary biologists have not been able to agree on a suitable definition of a species (Bush, 1982), though reproductive isolation of species serves as the ultimate yardstick, a yardstick missing in technological change. Thus it is hard to tell precisely a sailing ship from a steamship: the first steamships were really sailing ships with auxiliary steam engines; by the 1870s, they were steamships with auxiliary sails. The transition from one to the other was gradual enough to make the distinction arbitrary.

[69] The inclusion of the breastwheel may seem questionable, but the breastwheel was in truth a very different device than the under- and overshot waterwheel in use before. It combined the ease of construction and adaptability of the undershot waterwheel with the efficiency of the overshot wheel, and the water entered through a casing, which prevented spillage. On the other hand, John Rennie's sliding hatch, which allowed the water to enter the wheel at any height, depending on water supplies, must be viewed as a microinvention, as it improved the breastwheel but did not alter it in a radical way.

above, chlorine bleaching and ballooning fit this definition best. Moveable type and the weight-driven clock are also good examples. Not only do they embody some principle hitherto unknown or unused, but there is no gradual emergence of the new technique. Either it is there or it is not; suddenness is inevitable. The other type of macroinvention is the hybrid macroinvention, in which separate and previously known elements are combined together in a novel way. The biological equivalent is the mule or the nectarine, a combination of two different species combining the desirable properties of both. Crompton's mule was in fact called that for those very reasons. Another quintessential hybrid macroinvention was the windmill, combining elements of the water mill and the sailing ship. In hybrid macroinvention, the novelty is not so much the underlying idea itself, but realizing the potential of the combination of hitherto separate elements and the solution of the compatibility problems. In general, these macroinventions have tended to emerge somewhat more gradually than radical macroinventions. The breastwheel for instance (which combined the most desirable properties of the under- and the overshot wheel) appears for the first time in Jacob Leupold's *Theatrum Machinarum Generale* (published in 1724), but it was not until the brilliant engineer John Smeaton put his mind to perfecting them that breastwheels were recognized as a superior technology.

Why is this taxonomy useful? The answer is that microinventions and macroinventions follow different dynamics and respond to different incentives. Minor changes in existing technology are usually the result of systematic on-the-job tinkering and practical experimentation, and are thus a predictable by-product of normal productive activity. Firms employing a given technique will spend resources on refining it, adjusting it to its special needs and environment, and saving energy, labor, and raw material inputs all in order to reduce costs resulting in what economists have called localized technological change. It remains a stochastic process: many search efforts bear no result, whereas in other cases results are achieved with minimum effort. But by and large, microinventions follow the kind of rules that economists like Jacob Schmookler and Nathan Rosenberg have laid down. They respond to demand conditions, to changes in the prices of complements and substitutes, and in general differ only a little from other "inputs"—i.e., the firm or independent inventor will equate the last dollar spent on research and development to their expected marginal revenue product, as standard price theory suggests. Although invention is never "just another input," many discussions start from that point and then work out with great ingenuity the various complications arising from uncertainty, public good attributes, economies of scale, and the strategic behavior of innovative firms (Kamien and Schwartz, 1983).

What is not sufficiently realized is that this kind of technological change cannot persist in the long run without the periodical emergence of macroinventions. Continuous refinements of a given technology will eventually run into diminishing returns, and an economy in which all technical progress is of this type will find itself stagnating after a while. If microinventions were all there was to technological change, technological change as a source of economic growth would suffer from the same problem as Smithian economic growth: it can provide the economy with cheap lunches, but these lunches get increasingly skimpier and more expensive, and eventually some kind of equilibrium will be reached. This stationary state was the central intellectual achievement of classical political economy which never came to grips with macroinventions despite the classical political economists' understanding of the importance of technology. It is analogous to what evolutionary biologists refer to as stasis. The chief reason that there has been continuing technological progress in history is that these equilibria were punctuated by sudden changes of a bold and abrupt nature, followed by a wave of further improvements which eventually would peter out in the absence of subsequent microinventions.

Microinventions, the accumulation of small improvements, cannot alone account for the history of technology, and a continuous curve does not fit the observations. It is simply not the case that production can go from one technique to another by gradual, Darwinian change. Improve the axles, suspensions, couplings, and harness of a horse-and-buggy all you will, it will never become a bicycle. The history of engines is another example. There was much room for microinventions in the improvement of steam engines during the nineteenth century, and the gains from these inventions can readily be measured in terms of efficiency and compared with the gains from Watt's macroinvention (Von Tunzelmann (1981), p.

151). But those gains ran into diminishing returns as constraints dictated by the laws of thermodynamics were encountered. Continuous improvements in the steam engine did not lead smoothly to the Otto and Diesel engines; new breakthroughs were required. We are currently in a stage at which further improvements in internal combustion engines are running into diminishing returns. Another macroinvention may be necessary to resolve today's impasse in urban transportation. Sometimes even size alone made a difference: the power loom embodied principles similar to the ribbon loom perfected by Vaucanson in about 1760; but the difference in size was such that an entirely new solution was needed (Usher (1954), p. 284). Macroinventions can thus be thought of as shifting the marginal product curve of microinventive activity upward. The observation that feverish activity in improvement and refinement often followed major inventions is thus explained by this approach.

The economist's problem is that although we may be able to understand what led to one macroinvention or another, generalizing about them is all but impossible. At some level we stand in uncomprehending awe before the miracle of human genius. To borrow an example from another field: Franz Schubert wrote his Great C Major symphony in three weeks, limited only by the speed the quill could move on paper. It was neither performed nor published in his lifetime. Not even the most orthodox neoclassical economist will maintain seriously that cost-benefit calculations lend any insight into understanding why this uniquely perfect work came into existence. Similarly, the history of macroinventions is too complex to allow *any* generalization. In some cases, the breakthrough occurred gradually during many years of determined work and search, as in the case of Harrison's chronometer or the electric lightbulb. In others it was the result of a flash of inspiration. Watt's famous recounting of how the idea of the separate condenser occurred to him is worth citing:

> I had gone to take a walk on a fine Sabbath afternoon. I had entered the Green and passed the old washing house. I was thinking of the engine at the time. I had gone as far as the Lord's house when the idea came into my mind that . . . if a connection were made between the cylinder and an exhausting vessel it would rush into it and might there be condensed without cooling the cylinder . . . I had not walked farther than the Golf house, when the whole thing was arranged in my mind (cited by Usher, 1954, p. 71).

In still others, a lucky break was all that was needed, as in the case of Perkin's aniline dye or Goodyear's vulcanization of rubber—though in Goodyear's case success came after many years of frustrating search. However, the history of macroinventions supports Pasteur's observation that Fortune favors the prepared mind.

A variety of theories, the most famous of which is Usher's adaptation of *Gestalt* psychology, have been proposed to explain the emergence of major insights. Usher separated the act of invention into four stages: the perception of an incomplete pattern, the setting of the stage, the great insight, and the critical revision. Even in his analysis, the central event, that of "insight" remains somewhat mystical'

> How do new things happen? Is it possible to answer the question without invoking some external force, and without ... describing these activities of the mind as "unconscious" or "subconscious?" That an adequate answer is possible we may well believe; that we are now ready to answer the question admits of some doubt . . . It is not necessary to explain the final act of insight; the task now consists in explaining how the stage is set to suggest the solution of the perceived problem (Usher, 1954, p. 78).

Explaining technological change is thus a double task: we can explain reasonably well how a technique is improved and polished, but it is much harder to explain where it came from in the first place. Some simple-minded theories can be disposed of without too much ado. One of those is the modern variant of "necessity is the mother of invention," which is one of those oft-repeated cliches that also happen to be false.[70] This variant is expressed, for example, by A. E. Musson (1972, p. 42), who maintains that "there now seems little doubt that growing demand was the main driving force behind technological change." In the analysis of macroinventions, the symmetry between supply and demand breaks down. The reason is that for macroinventions, the social payoff is vastly larger than the social cost incurred in researching and developing them, so that there is always excess demand for them. An increase in demand would not make itself felt and makes little sense as a *primum*

[70] Cf. White (1978, p. 222): "Necessity is not the mother of invention, since all necessities are common to mankind living in similar natural environments . . . what seems needed and feasible to one culture may be a matter of indifference to another."

movens during any period of discontinuous technological change. To put it differently, macroinventions provide a free lunch; as long as the demand for goods and services is not satiated, the demand for technological change is by definition "there." It is always and everywhere supply constrained. To the extent that preferences and tastes mattered, they operated on the supply side. In societies in which the dominant attitudes were "If it ain't broke, don't fix it" or "If God had wanted us to fly, he would have given us wings," we would not expect much technological creativity, because there would be little "search" for inventions, and even accidental discoveries might not be accepted although the demand for goods produced by technology (and thus the derived demand for technology itself) is there. The limiting factors were, properly speaking, on the supply, not the demand side, although the distinction between the two here is somewhat tenuous. It is, of course, widely observed that increases in demand have accompanied technological change and Musson's view that the rise in demand for commodities *caused* invention seems consistent with that finding. But that inference is erroneous, because the increase in demand in periods of economic expansion such as the Industrial Revolution was itself primarily the *result* of rising incomes brought themselves about by rising productivity and technical progress, and thus the argument boils down to a logical circularity as technological change would be its own cause (Mokyr, 1985).

To be sure, demand for new technology can be defined in terms of what Landes refers to as bottlenecks, or Rosenberg's "focusing devices." Improvements that increased productivity in a complementary technique (e.g. spinning and weaving), or changes that raised the prices of substitutes (e.g. the exhaustion of sites for water mills near cheap labor), provided an incentive to search for technological solutions. Analogously, low wages are believed to have impeded technical progress, because primitive labor-intensive techniques remained profitable (Habakkuk, 1962). High wages allegedly stimulated the search for labor-saving inventions in general. The literature on this controversy is large and cannot be done justice to here. The weakness of this argument as an explanation of macroinventions is that what most macroinventions and many microinventions did was not to save one factor or another, but to save on *all* inputs, to reduce the pain and drudgery of work, and to improve the quality and durability of the goods produced. In other words, it provided a free lunch, and there is always excess demand for free lunches. Moreover, it was often impossible to know in advance whether an invention would be labor saving or capital saving and the ex post bias observed by the historian may well be different from the ex ante bias envisaged by the inventor, assuming he worried about factor proportions at all. As we consider inventions that were less and less pathbreaking and more and more "micro," the "free lunch" becomes smaller and smaller and we get closer and closer to a neoclassical world in which technology is just another input, governed by standard supply and demand mechanisms. Even then, the effect of changes in demand on the amount of technological progress depends on the supply curve. However, because nobody knows what is exactly on the price axis of this supply and demand diagram, it has limited explanatory power.

Explaining macroinventions requires more than supply and demand. Once more, the analogy with evolutionary biology can be enlightening. It has been known for a long time that the chances for speciation depend on certain environmental parameters. The parameters do not necessarily affect the chances of mutations occurring, but they determine the chances of the new species to survive and propagate. The most important of these parameters is that the new species be geographically separate from the main population, so as to allow it to interbreed (Mayr, 1970). There is no direct equivalent here in the growth of technologies, since the transmission of technological knowledge does not occur (usually) through sexual reproduction. But an appropriate environment for the propagation of macroinventions can readily be specified. It requires, above all, societies which allow original and skeptical minds to flourish. A society that does not tolerate deviants, eccentrics, and heterodox minds is not going to generate sudden breakthroughs necessary for macroinventions (Goldstone, 1987). Schumpeter (1934, p. 87) pointed out that resistance to innovation and novel ideas is never absent, and that overcoming it is "a special kind of task which does not exist in the customary course of life [and] requires a special conduct." The difficulty is that the ability to overcome hostility and the capacity to solve technical problems and make the new technique work were rarely concentrated in the same mind. Most deviants and eccentrics remain, of course, just that; they never invent anything while irritating and sometimes threatening the political and intellectual establishment. For quite obvious reasons, therefore, most societies have been less than tolerant to deviants. Yet societies, for reasons only dimly understood but which fall largely outside the realm of economics, have shifted back and forth in their willingness to put up with what they considered to be oddballs at best and heretics at worst. The correlation between pluralism and technological creativity is strong, but clearly not strong enough to indicate that it was either a sufficient or a necessary condition for technological progress.

A second set of environmental parameters which facilitate the emergence of macroinventions has to do with the institutional structure of societies, and in particular, with the specific institutions of capitalist, free-enterprise societies (Rosenberg and Birdzell, 1986). Even the unusual and original minds that sprouted ideas which eventually led to macroinventions needed minimum assurances, not only that they would not be persecuted for their "deviancy," but that society would in some way reward them for their effort.[71] Some technologically creative societies, like sixteenth century Holland and eighteenth and nineteenth century Britain, are associated with what might loosely be called capitalist institutions, but causality is hard to establish. The most plausible way of thinking about the association is that underlying social and intellectual currents create an atmosphere of rational materialism, which can give birth to both capitalist institutions and technological creativity, although these can be, and often have been, nipped in the bud by war, bad government, or natural disasters.

To some extent, the efficient enforcement of property rights and the absence of confiscatory taxation are necessary for *all* forms of economic development. But property rights in new information are complicated to enforce. Their principal form in Western societies is the patent system, yet the role of the patent system in furthering macroinventions is decidedly ambiguous. At first glance, it would appear that because most macroinventions generate very large

[71] Not all rewards had to be financial. Humphry Davy refused to patent the safety lamp, since he regarded it as a humanitarian device and wished to save as many lives as possible. However, he reputedly broke into a jealous fury when George Stephenson, who developed a similar device, claimed priority for his invention.

social surpluses at comparatively low costs, a patent system should be redundant, since it is enough for the inventor to capture a small proportion of the surplus for a brief period to make his efforts worthwhile. This would, overall, tend to impart a bias in favor of microinventions, since patents encourage relatively marginal inventions, in which the development costs are high or the expected payoff relatively low. Yet it is important to keep in mind that we are dealing with *expected* payoff. An invention could be "marginal" simply because the high payoff was coupled to a very low probability. In those cases, indeed, a patent system could have provided encouragement for macroinventions, although Scherer (1980, p. 448) believes that such cases were rare. Contemporaries of the Industrial Revolution believed that the patent system was essential (Watt wrote that "an engineer's life without patent was not worthwhile)," but the economic historian's verdict is decidedly mixed (Dutton, 1984).

The relation between the growth of commerce and that of technology is an interesting avenue toward understanding economic growth. Were highly commercialized societies more technologically creative than others? The last centuries suggest that this might have been the case: most societies which produced important innovations were mercantile and had a large commercial class that played an important role in technological changes. Yet medieval Europe, in which commerce had slowed down to a trickle, experienced a technological revival of dimensions unknown to Rome and Greece, which were by far more commercialized. Again, causality is difficult to establish. In some part, causality must have run in the other direction: technological progress reduced transportation and information costs and thus led to expanded trade. Trade, by establishing contact between previously separated cultures, led in its turn to technological changes through exposure effects. But on the most basic level, Smithian growth through gains from trade and Schumpeterian growth through the emergence of new ideas tended to be substitutes rather than complements. The Roman and Chinese empires were huge free trade zones in which institutions and investment projects favored trade, rather than technological progress. Schumpeter recognized that there was a trade-off between the conditions necessary for efficient and peaceful but more or less routinized exchange, permitting a regular free lunch through the gains from trade, and the feverish turbulence created by constant innovation.

An important explanation of technological creativity is that it is an autocorrelated process, so that it is explained, at least in part, by its own past. Theories which maintain that nothing succeeds like success and nothing fails like failure are easy to concoct. One view is that technological success creates a demonstration effect. By observing the careers of successful inventors such as Solvay or Edison, potential inventors learned that there were opportunities for riches and fame in successful inventing. Although the odds for success are very long, it may well be that most potential inventors systematically overestimate their own probability of success and thus overinvest in the search for new technologies. In this way, invention resembles professional athletics and lotteries. The enormous financial success of a few inspires vast efforts on the part of thousands to emulate them, despite a low probability of success. More important is the fact that knowledge generated by one technological breakthrough often spilled over onto other uses. The classic example is the invention of the mechanical clock, which created an entire industry of precision engineering in Renaissance Europe, producing instruments and toys requiring similar skills and relying on similar principles. As we have seen, the Jacquard loom, itself indebted to the binary logic of organs, inspired Charles Babbage's analytical engine (Babbage, 1864, pp. 117, 164) as well as Hollerith's use of punch cards to analyze the 1890 United States Census. German chemical knowledge, accumulated in producing dyes and fertilizer, was used with success in other chemical industries such as pharmaceutics and explosives.

The most widely accepted mechanism believed to be responsible for the autocorrelative pattern of technological change is that inventions created imbalances of factor and resource prices under the existing structure of allocation. As Rosenberg (1976, p. 111) puts it, "when processes are sufficiently complex and interdependent [they] involve an almost compulsive formulation of problems." The standard example is textile technology in the eighteenth century. The invention of the flying shuttle by Kay in 1733 increased the demand for yarn, because looms had become more efficient. Progress in technology resolved the spinning bottleneck, but by overshooting its mark, produced a new bottleneck in weaving.

When processes are complementary, technological progress in one stage will make it profitable to search for technological solutions in others. In many cases, of course, the complementarity was actually created by the invention: not until cheap steel could be made in the 1860s did it occur to anyone to produce steel ships and buildings. Rosenberg adds that relationships between components usually *had* to be corrected before an initial innovation could be fully exploited (1976, p. 117).

These "focusing devices" have substantial power in explaining the pattern of technological progress. They probably apply more to complex industrial technologies than, say, to agriculture. They increased the expected payoff from further search to resolve the bottleneck, but they do not guarantee it. The alternative to further invention is to reallocate resources toward the production stage that was not improved. Thus, inventing the flying shuttle created a bottleneck that could have been resolved by turning some weavers into spinners. In the nineteenth century, the demand for coal increased rapidly, creating a bottleneck in coal mining. Yet a satisfactory solution to the difficulties in applying inanimate power inside the mine was not found. The result was a massive reallocation of resources: the proportion of miners in the labor force in the second half of the nineteenth century doubled in Britain and almost tripled in the United States and Germany. The outcome depends on the supply side of technology, as well as on the cost of reallocation.[72] In any event, such compulsive sequences explain microinventions better than macroinventions, although some macroinventions (such as the Bessemer converter) can be analyzed in this way.

The main difficulty with theories that explain technological change by means of previous technological change is that they explain too much, not too little. It is not at all obvious that technological change is somehow self-sustaining. The evidence suggests that in fact, most technologically creative societies, have been so for limited time-spans.[73] Whether we look at the Nether-

[72] The same kind of objection holds for focusing devices resulting from negative supply shocks such as strikes, war, or depletion of natural resources.

[73] I have termed this phenomenon "Cardwell's Law" in honor of D. S. L. Cardwell (1972) who first pointed it out. For details, see Mokyr (1989b).

lands in its golden age, Britain during the Industrial Revolution, the United States in the twentieth century, or medieval Italy, it is clear that periods of technological creativity usually did not last very long. The mechanisms which account for positive autocorrelation, whatever they are, are soon dominated by more powerful forces which extinguish creativity and lead to stasis. In many cases the extinction of the creative spark has clearly identifiable causes: destructive wars, bad governments, and persecution of minorities can be clearly identified in some historical instances. In others, however, no obvious exogenous factors can be identified, and the weakening of progress had to be attributed to internal causes. Autocorrelation of technological creativity may thus have been less potent than is sometimes believed. In some cases, then, nothing succeeded as much as success; in others, nothing failed as much as success.

Analyzing the long-term historical record of technological progress suggests that it is difficult to identify factors either necessary or sufficient for technological creativity. The best we can do is to identify elements that could inhibit it: destructive civil wars and invasions, religious persecution, and prolonged epidemics and famines were incompatible with sustained progress. Apart from that, however, the principle of substitution obtained and there was usually more than one way not only to skin a cat, but also to make the skinning more efficient. Resource-rich societies found ways to create new technologies that used their resources; resource-poor societies found ways to save on expensive resources or invented ways to transport them cheaply. Societies with scarce labor produced labor-saving machinery, while societies with high population densities found ways to save and reclaim land from seas and deserts. Still other societies, with similar endowments of labor and resources, simply lived with their allocations and resigned themselves to a standard of living determined solely by population size and nature.

Bibliography

Arthur, Brian, "Competing Technologies and Lock-in by Historical Small Events: The Dynamics of Allocation under Increasing Returns," Stanford University, CEPR Publication No. 43 (1985).

Ashton, Thomas S., *Iron and Steel in the Industrial Revolution.* Manchester: Manchester University Press (1924).

Audin, Maurice, "Printing." In *A History of Technology and Invention* Vol. II: *The First Stages of Mechanization, 1450–1725,* edited by Maurice Daumas, pp. 620–667. New York: Crown (1969).

Audin, Maurice, "Printing" In *A History of Technology and Invention* Vol. III: *The Expansion of Mechanisation, 1725–1860*, edited by Maurice Daumas, pp. 656–705. New York: Crown (1979).

Babbage, Charles, *Passages from the life of a Philosopher,* London: Longman Green (1864).

Baldwin, William L. and Scott, John T., *Market Structure and Technological Change.* Chur, Switzerland: Harwood Academic Publishers (1987).

Barclay, Harold B., *The Role of the Horse in Man's Culture.* London: J. A. Allen (1980).

Bartley, W. W., "Philosophy of Biology versus Philosophy of Physics," in *Evolutionary Epistemology, Rationality, and the Sociology of Knowledge* edited by Gerard Radnitzky and W. W. Bartley, III, pp. 7–45. La Salle, Ill.: Open Court (1987).

Barraclough, K. C., *Steelmaking Before Bessemer,* 2 Vols. London: The Metals Society (1984).

Baumol, William J., "Entrepreneurship: Productive, Unproductive and Imitative; or the Rule of the Rules of the Game," Unpublished MS., Princeton University (1988).

Beaumont, Olga, "Agriculture: Farm Implements." in *A History of Technology,* Vol. IV: *The Industrial Revolution, 1750–1850,* edited by Charles Singer *et al.*, pp. 1–12. London: Oxford University Press (1958).

Bernal, J. D., *Science in History,* Vol. I: *The Emergence of Science.* Cambridge, MA: MIT Press (1965).

Blaine, Bradford B., "The Enigmatic Water Mill." In *On Pre-Modern Technology and Science: Studies in Honor of Lynn White,* edited by Bert S. Hall and Delno C. West. Malibu: Undena Publications, pp. 163–176 (1976).

Bloch, Marc, *Land and Work in Medieval Europe.* New York: Harper and Row (1966).

Boserup, E., *Population and Technological Change.* Chicago: University of Chicago Press (1981).

Boyd, Robert and Richerson, Peter J., *Culture and the Evolutionary Process.* Chicago: University of Chicago Press (1985).

Boulding, Kenneth, "Technology in the Evolutionary Process," in *The Trouble with Technology,* edited by S. McDonald, D. McL. Lamberton, T. Mandeville. New York: St. Martin's Press, pp. 4–10 (1983).

Bray, Francesca, "Agriculture". In *Science and Civilization in China,* edited by Joseph Needham, Vol. VI, Part II. Cambridge: Cambridge University Press, (1984).

Bray, Francesca, *The Rice Economies,* Oxford: Basil Blackwell (1986).

Brenner, Reuven, *History: The Human Gamble.* Chicago: University of Chicago Press (1983).

Bromehead, C. N., "Mining and Quarrying to the Seventeenth Century." In *A*

History of Technology, Vol. II: *The Mediterranean Civilizations and the Middle Ages, 700 B.C. to A.D. 1500*, edited by Charles Singer *et al.*, Vol. II, pp. 1–40. London: Oxford University Press (1956).
Brown, Shannon R., "The Ewo Filature: a Study in the Transfer of Technology to China," *Technology and Culture,* Vol. 20, No. 3 (July), pp. 550–568 (1979).
Brumbaugh, Robert S., *Ancient Greek Gadgets and Machines*. New York: Thomas Y. Crowell (1966).
Bryant, Lynwood, "The Silent Otto," *Technology and Culture,* Vol. 7, No. 2 (Spring), pp. 184–200 (1966).
Bryant, Lynwood, "The Beginnings of the Internal Combustion Engine," in *Technology in Western Civilization,* edited by Melvin Kranzberg and Carroll W. Pursell, Jr., Vol. I, pp. 648–663. N.Y.: Oxford University Press (1967).
Bryant, Lynwood, "Rudolf Diesel and his Rational Engine," *Scientific American,* Vol. 221, No. 2 (Aug), pp. 108–117 (1969).
Bryant, Lynwood, "The Role of Thermodynamics: the Evolution of the Heat Engine," *Technology and Culture,* Vol. 14, No. 2 (part I), pp. 152–165 (1973).
Buchanan, R. A., "The Diaspora of British Engineering," *Technology and Culture,* Vol. 27, No. 3 (July), pp. 501–24 (1986).
Bulliet, Richard W., *The Camel and the Wheel.* Cambridge: Harvard University Press (1975).
Burstall, A. F., *A History of Mechanical Engineering*. Cambridge, Mass: MIT Press (1965).
Bush, Guy L., "What Do We Really Know About Speciation?" in *Perspectives on Evolution,* edited by Roger Milkman. Sunderland, Mass: Sinauer Publishing Co, pp. 119–28 (1982).
Campbell, Donald T., "Blind Variation and Selective Retention in Creative Thought as in Other Knowledge Processes," in *Evolutionary Epistemology, Rationality, and the Sociology of Knowledge* edited by Gerard Radnitzky and W. W. Bartley, III, pp. 91–114, La Salle, Ill: Open Court (orig. publ. in 1960) (1987).
Cardwell, D. S. L., *From Watt to Clausius: The Rise of Thermodynamics in the Early Industrial Age*. Ithaca: Cornell University Press (1971).
Cardwell, D. S. L., *Turning Points in Western Technology.* New York: Neale Watson (1972).
Casson, L., *Ships and Seamanship in the Ancient World.* Princeton: Princeton University Press (1971).
Chao, Kang, *The Development of Cotton Textile Production in China.* Cambridge, Mass.: Harvard University Press (1977).
Chao, Kang, *Man and Land in Chinese History: An Economic Analysis.* Stanford: Stanford University Press (1986).
Chapman, S. D., *The Cotton Industry in the Industrial Revolution.* London: Macmillan (1972).
Chatterton, E. Keble, *Sailing Ships and their Story,* London: Sidgwick and Jackson (1909).
Cipolla, Carlo M., *The Economic History of World Population,* Harmondsworth, Pelican Books (1965a).
Cipolla, Carlo M., *Guns and Sails in the Early Phase of European Expansion, 1400–1700*, London: Collins (1965b).
Cipolla, Carlo M., *Clocks and Culture, 1300–1700*, New York: Norton (1967).
Cipolla, Carlo M., *Literacy and Development in the West,* Harmondsworth: Penguin Books (1969).
Clapham, Michael, "Printing," In *A History of Technology,* Vol. III: *From the Renaissance to the Industrial Revolution, 1500–1750,* edited by Charles Singer *et*

al., pp. 377–410. London: Oxford University Press (1957).
Clow, Archibald and Clow, Nan L., *The Chemical Revolution*, London: The Batchworth Press (1958).
Clow, Archibald and Nan L., "The Chemical Industry: Interaction with the Industrial Revolution." In *A History of Technology*, Vol. IV: *The Industrial Revolution, 1750–1850*, edited by Charles Singer *et al.*, pp. 230–257. London: Oxford University Press (1958a).
Clow, Archibald and Nan L., "Ceramics from the Fifteenth Century to the Rise of the Staffordshire Potteries." In *A History of Technology*, Vol. IV: *The Industrial Revolution, 1750–1850*, edited by Charles Singer *et al.*, pp. 328–357. London: Oxford University Press (1958b).
Coombs, Rod, Paolo Saviotti and Vivien Walsh, *Economics and Technological Change*. Totowa, N.J.: Rowman and Littlefield (1987).
Cooper, Carolyn, "The Portsmouth System of Manufacture," *Technology and Culture*, **25**, No. 2 (April), pp. 182–225 (1984).
Crosby, Alfred, *The Columbian Exchange*; *Biological and Cultural Consequences of 1492*. Westport, Conn., Greenwood Publishing Co. (1972).
Crosby, Alfred, *Ecological Imperialism*: *The Biological Expansion of Europe, 900–1900*, Cambridge: Cambridge University Press (1986).
Crouzet, François, *The First Industrialist*, Cambridge; Cambridge University Press (1985).
Cutting, C. L., "Fish Preservation." In *A History of Technology*, Vol. IV: *The Industrial Revolution, 1750–1850*, edited by Charles Singer *et al.*, pp. 44–55. London: Oxford University Press (1985).
Daumas, Maurice and Paul Gille, "Methods of Producing Power." In *A History of Technology and Invention*, Vol. III: *The Expansion of Mechanization*, 1725–1860, edited by Maurice Daumas, pp. 17–80. New York: Crown (1979).
Daumas, Maurice, "The Extraction of Chemical Products." In *A History of Technology and Invention*, Vol. II: *The First Stages of Mechanization, 1450–1725*, edited by Maurice Daumas, pp. 172–211. New York: Crown (1964).
David, P. A., *Technical Choice, Innovation and Economic Growth*. Cambridge; Cambridge University Press (1975).
David, P. A., "The Hero and the Herd in Technological History: Reflections on Thomas Edison and the 'Battle of the Systems'." Unpublished paper, Stanford University (1987).
Davis, Lance E., Robert G. Gallman and T. D. Hutchins, "Call me Ishmael not Domingo Floresta: The Rise and Fall of the American Whaling Industry," in *The Vital One*: *Essays Presented to Jonathan R. T. Hughes*, edited by Joel Mokyr. Greenwich, CT.: JAI Press (1989).
Derry, T. K. and T. I. Williams, *A Short History of Technology*, Oxford: Oxford University Press (1960).
Deshayes, Jean, "Greek Technology." *In a History of Technology and Invention*, Vol. I: *The Origins of Technological Civilization*, edited by Maurice Daumas, pp. 181–214, New York: Crown (1969).
Dickinson, H. W., "The Steam-Engine to 1830." In *A History of Technology*, Vol. IV: *The Industrial Revolution, 1750–1850*, edited by Charles Singer *et al.*, pp. 168–198. London: Oxford University Press (1958).
Dosi, Giovanni, *Technical Change amd Industrial Transformation*. New York: St. Martin's (1984).
Dresbeck, LeRoy, "Winter Climate and Society in the Northern Middle Ages: the Technological Impact," in *On Pre-Modern Technology and Science*: *Studies in*

Honor of Lynn White, edited by Bert S. Hall and Delno C. West. Malibu: Undena Publications, pp. 177–199.

Dutton, H. I., *The Patent System and Inventive Activity during the Industrial Revolution*. Manchester: Manchester University Press (1984).

Duval, Paul-Marie, "The Roman Contribution to Technology," In *A History of Technology and Invention*, Vol. I: *The Origins of Technological Civilization*, edited by Maurice Daumas, pp. 216–259, New York: Crown (1962).

Eldredge, Niles, *Time Frames: The Rethinking of Darwinian Evolution and the Theory of Punctuated Equilibria*. New York: Simon and Schuster (1985).

Elster, Jon, *Explaining Technical Change*, Cambridge: Cambridge University Press (1983).

Elton, Sir Arthur, "Gas for Light and Heat." In *A History of Technology*, Vol. IV: *The Industrial Revolution, 1750–1850*, edited by Charles Singer *et al.*, pp. 258–276. London: Oxford University Press (1958).

Elvin, M., *The Pattern of the Chinese Past*. Stanford: Stanford University Press (1973).

English, W., "The Textile Industry: Silk Production and Manufacture, 1750–1900." In *A History of Technology*, Vol. IV: *The Industrial Revolution, 1750–1850*, edited by Charles Singer *et al.*, pp. 308–327 London: Oxford University Press (1958).

Eudier, Walter and Jacques Payen, "The Spinning of Textible Fibers." In *A History of Technology and Invention*, Vol. III: *The Expansion of Mechanization, 1725–1860*, edited by Maurice Daumas, pp. 583–598. New York: Crown (1979).

Evans, Francis T., "Roads, Railways, and Canals: Technical Choices in Nineteenth Century Britain," *Technology and Culture*, Vol. 22, No. 1 (January), pp. 1–34 (1981).

Fenoaltea, Stefano, "Slavery and Supervision in Comparative Perspective: a Model," *Journal of Economic History*, **44**, No. 3 (Sept), pp. 635–68 (1984).

Ferguson, Eugene S., "History and Historiography," in *Yankee Enterprise: The Rise of the American System of Manufactures*, edited by Otto Mayr and Robert C. Post. Washington: the Smithsonian Institution, pp. 1–23 (1981).

Feuerwerker, A., "The State and the Economy in Late Imperial China," *Theory and Society*, Vol 13, pp. 297–326 (1984).

Finley, M. I., "Technical Innovation and Economic Progress in the Ancient World," *Economic History Review*, second series, Vol. XVIII, No. 1 (Aug), pp. 29–45 (1965).

Finley, M. I., *The Ancient Economy*, Berkeley: University of California Press (1973).

Fletcher, R. A., *Steam-Ships: The Story of their Development to the Present Day*, Philadelphia: J. B. Lippincott (1910).

Foley, Vernard, "Leonardo, the Wheel Lock, and the Milling Process," *Technology and Culture*, Vol. 24, No. 3 (July), pp. 399–427 (1983).

Forbes, R. J., "Metallurgy." In *A History of Technology*, Vol. II: *The Mediterranean Civilizations and the Middle Ages, 700 B.C. to A.D. 1500*, edited by Charles Singer *et al.*, pp. 41–80. London: Oxford University Press (1956a).

Forbes, R. J., "Food and Drink." In *A History of Technology*, Vol. II: *The Mediterranean Civilizations and the Middle Ages, 700 B.C. to A.D. 1500*, edited by Charles Singer *et al.*, pp. 103–146. London: Oxford University Press (1956b).

Forbes, R. J., "Power." In *A History of Technology*, Vol. II: *The Mediterranean Civilizations and the Middle Ages, 700 B.C. to A.D. 1500*, edited by Charles Singer *et al.*, pp. 589–628. London: Oxford University Press (1956c).

Forbes, R. J., "Food and Drink." In *A History of Technology*, Vol. III: *From the*

Renaissance to the Industrial Revolution, 1500–1750, edited by Charles Singer *et al.*, pp. 1–26. London: Oxford University Press (1957).

Forbes, R. J., "Power to 1850." In *A History of Technology*, Vol. IV: *The Industrial Revolution, 1750–1850,* edited by Charles Singer *et al.*, pp. 148–167. London: Oxford University Press (1958a).

Forbes, R. J., *Man the Maker*. London: Abelard-Shuman Ltd. (1958b).

Fossier, Rober, *Enfance de l'Europe,* 2 vols. Paris: Presses Universitaires de France (1982).

Friedel, Robert, "Parkesine and Celluloid: The Failure and Success of the First Modern Plastic," *History of Technology,* Vol. 4, pp. 45–62 (1979).

Fussell, G. E., "Agriculture: Techniques of Farming." In *A History of Technology,* Vol. IV: *The Industrial Revolution, 1750–1850*, edited by Charles Singer *et al.*, pp. 13–43. London: Oxford University Press (1958).

Futia, C. A., "Schumpeterian Competition," *Quarterly Journal of Economics,* Vol. 94, pp. 675–95 (1980).

Gernet, Jacques, *A History of Chinese Civilization.* Cambridge: Cambridge University Press (1982).

Gernsheim, Helmut and Alison, "The Photographic Arts: Photography." In *A History of Technology,* Vol. V: *The Late Nineteenth Century,* edited by Charles Singer *et al.*, pp. 716–733. London: Oxford University Press (1958).

Gibbs, F. W., "Invention in Chemical Industries." In *A History of Technology,* Vol. III: *From the Renaissance to the Industrial Revolution, 1500–1750*, edited by Charles Singer *et al.*, pp. 676–708. London: Oxford University Press (1957).

Gibbs, F. W., "Extraction and Production of Metals." In *A History of Technology,* Vol. IV: *The Industrial Revolution, 1750–1850*, edited by Charles Singer *et al.*, pp. 118–147. London: Oxford University Press (1958).

Gibbs, F. W., "Invention in Chemical Industries." In *A History of Technology,* Vol. III: *From the Renaissance to the Industrial Revolution, 1500–1750*, edited by Charles Singer *et al.*, pp. 676–708. London: Oxford University Press (1957).

Giedion, Siegfried, *Mechanization Takes Command.* New York: W. W. Norton (1948).

Gilbert, K. R., "Machine-Tools." In *A History of Technology,* Vol. IV: *The Industrial Revolution, 1750–1850,* edited by Charles Singer *et al.*, pp. 417–441. London: Oxford University Press (1958).

Gilfillan, S. C., *The Sociology of Invention.* Cambridge, Mass.: M.I.T. Press (1935).

Gille, Bertrand, "Machines." In *A History of Technology,* Vol. II: *The Mediterranean Civilizations and the Middle Ages,* 700 *B.C. to A.D.* 1500, edited by Charles Singer *et al.*, pp. 629–657. London: Oxford University Press (1956).

Gille, Bertrand, "The Medieval Age of the West, Fifth Century to 1350." In *A History of Technology and Invention,* Vol. I: *The Origins of Technological Civilization,* edited by Maurice Daumas, Vol. I, pp. 422–572. New York: Crown (1962).

Gille, Bertrand, *Engineers of the Renaissance.* Cambridge Mass.: M.I.T. Press (1966).

Gille, Bertrand, "The Fifteenth and Sixteenth Centuries in the Western World." In *A History of Technology and Invention,* Vol. II: *The First Stages of Mechanization, 1450–1725*, edited by Maurice Daumas, pp. 16–148. New York: Crown (1969).

Gille, Bertrand, *Histoire des Techniques*: *Technique et Civilisations, Technique et Sciences.* Paris: Editions Gallimard, (1978).

Gille, Bertrand, "The Evolution of Metallurgy." In *A History of Technology and Invention,* Vol. III: *The Expansion of Mechanization, 1725–1860*, edited by Maurice Daumas, pp. 527–553. New York: Crown (1979).

Gille, Paul, "From the Traditional Methods to Steam," In *A History of Technology* and Invention, Vol. II: *The First Stages of Mechanization, 1450–1725*, edited by Maurice Daumas, pp. 437–463. New York: Crown (1969a).

Gille, Paul, "Sea and River Transportation." In *A History of Technology and Invention,* Vol. II: *The First Stages of Mechanization, 1450–1725*, edited by Maurice Daumas, pp. 361–436. New York: Crown (1969b).

Gillispie, Charles, *The Montgolfier Brothers and the Invention of Aviation,* Princeton: Princeton. University Press (1983).

Gimpel, J., *The Medieval Machine: The Industrial Revolution of the Middle Ages.* Harmondsworth: Penguin Books (1976).

Golas, P. J., "Chinese Mining: Where Was the Gunpowder?" In *Explorations in the History of Science and Technology in China,* edited by Li Guohao, Zhang Mengwen, and Cao Tianqin, pp. 453–458. Shanghai: Chinese Classics Publishing House (1982).

Goldschmidt, Richard, *The Material Basis of Evolution,* New Haven: Yale University Press (1940).

Goldstone, Jack A., "Geopolitics, Cultural Orthodoxy, and Innovation," *Sociological Theory,* Vol. 5, No. 2 (Fall 1987), pp. 119–135 (1987).

Gordon, Robert B., "Hydrological Science and the Development of Waterpower for Manufacturing," *Technology and Culture,* Vol. 26, No. 2 (April), pp. 204–235 (1985).

Gould, Stephen Jay, "The Meaning of Punctuated Equilibrium and its Role in Validating a Hierarchial Approach to Macroevolution," in *Perspectives on Evolution,* edited by Roger Milkman. Sunderland, Mass.: Sinauer Publishing Co, pp. 83–104 (1982).

Guohao, L., Z. Mengwen and C. Tianqin, *Explorations in the History of Science and Technology in China.* Shanghai: Chinese Classics Publishing House (1982).

Habakkuk, H. J., *American and British Technology in the Nineteenth Century.* Cambridge: Cambridge University Press (1962).

Haber, L. F., *The Chemical Industry during the Nineteenth Century.* Oxford: At the Clarendon Press (1958).

Hacker, Barton, "Greek Catapults and Catapult Technology: Science, Technology and War in the Ancient World," *Technology and Culture,* Vol. 9, No. 1 (January), pp. 34–50 (1968).

Hacker, Barton, "The Weapons of the West: Military Technology and Modernization in Nineteenth Century China and Japan," *Technology and Culture,* Vol. 18, No. 1 (January), pp. 43–55 (1977).

Hall, A. Rupert, "Early Modern Technology to 1600, *Technology in Western Civilization,* edited by Melvin Kranzberg and Carroll W. Pursell, Jr., Vol. I, pp. 79–106, N.Y.: Oxford University Press (1967).

Hall, A. Rupert, "On Knowing and Knowing how to . . .," *History of Technology,* Vol. III, Pp. 91–104 (1978).

Hall, Marie Boas, "The Strange Case of Aluminium," *History of Technology,* Vol. I, pp. 143–157 (1976).

Harley, C. Knick, "On the Persistence of Old Techniques: The Case of North American Wooden Shipbuilding," *Journal of Economic History,* Vol. XXXIII, No. 2 (June), pp. 372–389 (1973).

Harms, Robert, *Games Against Nature.* Cambridge: Cambridge University Press (1987).

Harrison, J. A., *The Chinese Empire.* New York: Harcourt Brace (1972).

Hartwell, Robert M., "Historical Analogism, Public Policy, and Social Science in Eleventh- and Twelfth Century China," *Amerian Historical Review,* Vol. 76, No. 3 (June), pp. 690–727 (1971).

al-Hassan, Ahmad Y. and Hill, Donald R., *Islamic Technology*. Cambridge: Cambridge University Press (1986).

Headrck, Daniel, *The Toolsof Empire*. New York: Oxford University Press (1981).

Heertje, Arnold, "Can We Explain Technical Change," in *The Trouble with Technology,* edited by S. McDonald, D. McL. Lamberton and T. Mandeville. New York: St. Martin's Press, pp. 37–49 (1983).

Hicks, J. R., *A Theory of Economic History*. London: Oxford University Press (1969).

Hill, Donald R., "The Banu Musà and their 'Book of Ingenious Devices'," *History of Technology,* Vol II, pp. 39–76 (1977).

Hill, Donald R., "Information on Engineering in the Works of Muslim Geographers," History of Technology, Vol. IX, pp. 127–142 (1984a).

Hill, Donald R., *A History of Engineering in Classical and Medieval Times*. London: Croom Helm (1984b).

Hills, Richard L., *Power in the Industrial Revolution*. Manchester: Manchester University Press (1970).

Hills, Richard, L., "Hargreaves, Arkwright, and Crompton: Why Three Separate Inventors?," *Textile History,* Vol. 10, pp. 114–26 (1979).

Hindle, Brooke, *Emulation and Invention*. New York: New York University Press (1981).

Hodges, H., *Technology in the Ancient World*. London: Allen Lane, The Penguin Press (1970).

Hohenberg, Paul, *Chemicals in Western Europe, 1850–1914*. Chicago: Rand McNally (1976).

Hollister-Short, G. J., "Leads and Lags in Late Seventeenth Century English Technology," *History of Technology,* Vol. I, pp. 159–183 (1976).

Hollister-Short, G. J., "Gunpowder and Mining in Sixteenth- and Seventeenth Century Europe," *History of Technology,* Vol. 10, pp. 31–66 (1985).

Holmyard, E. J., "The Chemical Industry: Developments in Chemical Theory and Practice." In *A History of Technology,* Vol. IV: *The Industrial Revolution, 1750–1850,* edited by Charles Singer *et al.* pp. 214–229. London: Oxford University Press (1985a).

Holmyard, E. J., "Dyestuffs in the Nineteenth Century." In *A History of Technology,* Vol. V: *The Late Nineteenth Century,* edited by Charles Singer *et al.,* pp. 257–283. London: Oxford University Press (1985b).

Hounshell, David A., "Elisha Gray and the Telephone," *Technology and Culture,* Vol. 16, No. 2 (April), pp. 133–61 (1975).

Hounshell, David A., *From the American System to Mass Production,* 1800–1932. Baltimore: The Johns Hopkins Press. (1984).

Howard, Robert, "Interchangeable Parts Reexamined: The Private Sector of the American Arms Industry on the Eve of the Civil War," *Technology and Culture,* Vol. 19, No. 4 (October), pp. 633–49 (1978).

Huang, Philip C. C., 1985. *The Peasant Economy and Social Change in North China.* Stanford: Stanford University Press (1985).

Huard, Pierre *et al.* "The Techniques of the Ancient Far East." In *A History of Technology and Invention,* Vol. I: *The Origins of Technological Civilization,* edited by Maurice Daumas, pp. 216–259. New York: Crown (1969).

Hucker, Charles O., *China's Imperial Past.* Stanford: Stanford University Press (1975).

Hughes, Jonathan R. T., *The Vital Few.* New York: Oxford University Press, second ed. (1986).

Hughes, J. Donald. *Ecology in Ancient Civilizations.* Albuquerque: University of

New Mexico Press (1975).

Hull, David L., "A Mechanism and its Metaphysics: an Evolutionary Account of the Social and Conceptual Development of Science," *Biology and Philosophy*. (1988), forthcoming.

Hunter, Dard, *Papermaking through Eighteen Centuries*. New York: William Edwin Rudge (1930).

Jewkes, J., Sawers, D. and R. Stillerman, *The Sources of Invention*. 2nd ed. New York: Norton (1969).

Jones, Eric L., *The European Miracle*. Cambridge: Cambridge University Press (1981).

Jones, Eric L., *Growth Recurring: Economic Change in World History*. Oxford: The Clarendon Press (1988).

Jope, E. M., "Vehicles and Harness," In *A History of Technology*, Vol. II: *The Mediterranean Civilizations and the Middle Ages, 700 B.C. to A.D. 1500*, edited by Charles Singer *et al.*, pp. 537–562. London: Oxford University Press (1956a).

Jope, E. M., "Agricultural Implements." In *A History of Technology*, Vol. II: *The Mediterranean Civilizations and the Middle Ages, 700 B.C. to A.D. 1500*, edited by Charles Singer *et al.*, pp. 81–102. London: Oxford University Press (1956b).

Kahn, Arthur D., "Greek Tragedians and Science and Technology," *Technology and Culture*, Vol. 11 No. 2 (April), pp. 133–62 (1970).

Kamien, Morton I and Schwartz, Nancy L., *Market Structure and Innovation*. Cambridge: Cambridge University Press (1982).

Kanefsky, J. and J. Robey, "Steam Engines in 18th Century Britain: A Quantitative Assessment," *Technology and Culture*, Vol. 21, No. 2, April, pp. 161–186 (1980).

Kaufer, Erich, *The Economics of the Patent System*. Chur, Switzerland: Harwood Academic Publishers (1989).

Kreutz, Barbara M., "Mediterranean Contributions to the Medieval Mariner's Compass," *Technology and Culture*, Vol. 14, No. 3 (July), pp. 367–383 (1973).

Lach, Donald F., *Asia in the Making of Europe*, Vol. II: *A Century of Wonder*. Chicago: University of Chicago Press (1977).

Landels, J. G., *Engineering in the Ancient World*. Berkeley: University of California Press (1978).

Landes, D., *The Unbound Prometheus*. Cambridge: Cambridge University Press (1969).

Landes, D., *Revolution in Time*. Cambridge, Mass.: Harvard University Press (1983).

Lane, Frederic C., "The Economic Meaning of the Compass," *American Historical Review*, Vol. LXVIII, No. 3 (April), pp. 605–617 (1963).

Lazonick, William, "Production Relations, Labor Productivity, and Choice of Technique: British and U.S. Spinning," *Journal of Economic History*, Vol. XLI, No. 3 (Sept 1981), pp. 491–516 (1981).

Lee, Desmond, "Science, Philosophy, and Technology in the Greco-Roman World", *Greece and Rome*, Second Series, Vol. XX, No. 1 (April), pp. 65–78 and No. 2 (Oct), pp. 180–193 (1973).

LeGoff, Jacques, *Time, Work, and Culture in the Middle Ages*. Chicago: University of Chicago Press (1980).

Lethbridge, T. C., "Shipbuilding." In *A History of Technology*, Vol. II: *The Mediterranean Civilizations and Middle Ages, 700 B.C. to A.D. 1500*, edited by

Charles Singer *et al.*, pp. 563–588. London: Oxford University Press (1956).

Lewis, Bernard, *The Muslim Discovery of Europe*. New York: Norton (1982).

Lilley, S., *Men, Machines, and History*. New York: International Publishers (1965).

Machabey, Armand, "Techniques of Measurement." In *A History of Technology and Invention,* Vol. II: *The First Stages of Mechanization, 1450–1725,* edited by Maurice Daumas, pp. 306–343. New York: Crown (1969).

Mann, Julia De L., "The Textile Industry: Machinery for Cotton, Flax, Wool, 1760–1900." In *A History of Technology,* Vol. IV: *The Industrial Revolution, 1750–1850,* edited by Charles Singer *et al.*, pp. 277–307. London: Oxford University Press (1958).

Mantoux, Paul, *The Industrial Revolution in the Eighteenth Century*. New York: Harper and Row (orig. published in 1905) (1961).

Manuel, F., "The Luddite Movement in France," *Journal of Modern History,* Vol. 10, No. 2, pp. 180–211 (1938).

Marshall, Alfred, *Industry and Trade*. London: MacMillan and Co. (1919).

Mayr, Ernst, *Population, Species, and Evolution*. Cambridge, Mass.: The Belknap Press (1970).

Mayr, Otto, "The Science-Technology Relationship as a Historiographical Problem," *Technology and Culture,* Vol. 17 No. 4 (October), pp. 663–72 (1976).

McCloskey, Donald N., "The Industrial Revolution 1780–1860: a survey," in *The Economic History of Britain since 1700,* edited by Roderick Floud and Donald N. McCloskey, Vol. I, pp. 103–127, Cambridge: Cambridge University Press (1981).

McCloy, Shelby T., *French Inventions of the Eighteenth Century*. Kentucky: University of Kentucky Press (1952).

McNeill, W. H., *The Pursuit of Power*. Chicago: University of Chicago Press (1982).

Merton, Robert K., *The Sociology of Science*. Chicago: University of Chicago Press (1973).

Metzger, Thomas, "On the Historical Roots of Economic Modernization in China: The Increasing Differentiation of the Economy from the Polity During Late Ming and Early C'hing Times," in *Modern Chinese Economic History* ed. by C. Hou and T. Yu, pp. 3–14. Taipei: Institute of Economics, Academia Sinica (1979).

Miller, Harry, "Potash from Wood Ashes: Frontier Technology in Canada and the U.S.," *Technology and Culture,* Vol. 21, No. 2 (April), pp. 187–208 (1980).

Minchinton, W. E., "Early Tide Mills: Some Problems," *Technology and Culture,* Vol. 20, No. 4 (October), pp. 777–786 (1979).

Mokyr, Joel, ed., *The Economics of the Industrial Revolution*. Totowa, New Jersey: Rowman and Allanheld (1985).

Mokyr, Joel, "Was There a British Industrial Evolution?" in *The Vital One: Essays Presented to Jonathan R. T. Hughes,* edited by Joel Mokyr. Greenwich, CT.: JAI Press (1989a).

Mokyr, Joel, *The Lever of Riches: Technological Creativity and Economic Progress*. New York, Oxford University Press, (1989b).

Morison, Elting E., *Men, Machines, and Modern Times*. Cambridge, Mass., M.I.T. Press (1966).

Morris, T. N., "Management and Preservation of Food," In *A History of Technology,* Vol. V: *The Late Nineteenth Century,* edited by Charles Singer *et al.*, pp. 26–52. London: Oxford University Press (1958).

Mowery, David, "Industrial Research, 1900–1950," in *The Decline of the British Economy,* edited by Bernard Elbaum and William Lazonick. Oxford: Clarendon Press, pp. 189–222.

Multhauf, Robert P., "Industrial Chemistry in the Nineteenth Century", in *Technology in Western Civilization,* edited by Melvin Kranzberg and Carroll W. Pursell, Jr., Vol. I, pp. 468–488. N.Y.: Oxford University Press (1967).
Multhauf, Robert P., "The French Crash Program for Saltpeter Production, 1776–94", *Technology and Culture,* Vol. 12 No. 2 (April), pp. 163–181 (1971).
Mumford, Lewis. *Technics and Civilization.* New York, Harcourt, Brace & World (orig published in 1934) (1963).
Musson, A. E., *Science, Technology and Economic Growth in the Eighteenth Century.* London: Methuen (1972).
Musson, A. E., "Joseph Whitworth and the Growth of Mass-Production Engineering," *Business History,* Vol. XVII No. 2 (July), pp. 109–49 (1975a).
Musson, A. E., "Continental Influences on the Industrial Revolution in Great Britain," in *Great Britain and Her World: Essays in Honor of W. O. Henderson,* edited by Barrie M. Ratcliffe. Manchester: Manchester University Press (1975b).
Musson, A. E., "British Origins," in *Yankee Enterprise: The Rise of the American System of Manufactures,* edited by Otto Mayr and Robert C. Post. Washington: the Smithsonian Institution, pp. 25–48.
Musson, A. E. and Eric Robinson, *Science and Technology in the Industrial Revolution.* Manchester: Manchester University Press (1969).
Naish, George, "Ship-Building." In *A History of Technology,* Vol. IV: *The Industrial Revolution, 1750–1850,* edited by Charles Singer *et al.,* pp. 574–595. London: Oxford University Press (1958).
Needham, J., *The Development of Iron and Steel Technology in China.* Cambridge; W. Heffer (1964).
Needham, J., *Science and Civilization in China,* Vol. 4 (Physical and Physical Technology), part 2: *Mechanical Engineering.* Cambridge: Cambridge University Press (1965).
Needham, J., *The Grand Titration.* Toronto: Toronto University Press (1969).
Needham, J., *Clerks and Craftsmen in China and the West.* Cambridge: Cambridge University Press (1970).
Needham, J., "History and Human Values: A Chinese Perspective for World Science and Technology." *Occasional Papers, Centre for East Asian Studies, McGill University,* Montreal (1975).
Needham, J., *Science in Traditional China.* Cambridge, Mass.: Harvard University Press (1981).
Nef, John U., *War and Human Progress.* Cambridge: Harvard University Press (1950).
Nef, John U., "Coal Mining and Utilization." In *A History of Technology,* Vol. III: *From the Renaissance to the Industrial Revolution, 1500–1750,* edited by Charles Singer *et al.,* pp. 72–88. London: Oxford University Press (1957).
Nef, John U., *The Conquest of the Material World.* Chicago: University of Chicago Press (1964).
Nelson, Richard R. and Winter, Sidney, *An Evolutionary Theory of Economic Change.* Cambridge; The Belknap Press (1982).
North D. C. and R. P. Thomas, *The Rise of the Western World.* Cambridge; Cambridge University Press (1973).
North, D. C., *Structure and Change in Economic History.* New York: W. W. Norton (1981).
Oleson, John Peter, *Greek and Roman Mechanical Water-Lifting Devices: The History of a Technology.* Toronto: University of Toronto Press (1984).
Olmstead, Alan, "The Mechanization of Reaping and Mowing in American

Agriculture, 1833–1870." *Journal of Economic History,* Vol. XXX, No. XXXV, No. 2 (June).

Olson, Mancur, *The Rise and Decline of Nations.* New Haven: Yale University Press (1982).

Ovitt, George, Jr., "The Cultural Context of Western Technology: Early Christian Attitudes toward Manual Labor," *Technology and Culture,* Vol. 27 No. 3 (July), pp. 477–500 (1986).

Pacey, Arnold, *The Maze of Ingenuity: Ideas and idealism in the Development of Technology.* New York: Holmes and Meier (1975).

Parker, W., *Europe, America, and the Wider World.* Cambridge: Cambridge University Press (1984).

Parry, J. H., *The Discovery of the Sea.* Berkeley: University of California Press (1974).

Patterson, R., "Spinning and Weaving." In *A History of Technology,* Vol. II: *The Mediterranean Civilizations and the Middle Ages, 700 B.C. to A.D. 1500,* edited by Charles Singer *et al.*, pp. 191–220. London: Oxford University Press (1956).

Patterson, R., "Spinning and Weaving." In *A History of Technology,* Vol. III: *From the Renaissance to the Industrial Revolution, 1500–1750,* edited by Charles Singer *et al.*, pp. 151–180. London: Oxford University Press (1957).

Paulinyi, Akos, "Revolution and Technology," in *Revolution in History,* edited by Roy Porter and Mikulas Teich. Cambridge: Cambridge University Press (1986).

Payen, Jacques and Jean Pilisi, "Weaving and Mechanical Finishing." In *A History of Technology and Invention,* Vol. III: *The Expansion of Mechanization, 1725–1860,* edited by Maurice Daumas, pp. 599–618. New York: Crown (1979).

Perkin, H. J., *The Origins of Modern English Society, 1780–1880.* London (1969).

Perkins, Dwight H., "Government as an Obstacle to Industrialization: The Case of Nineteenth Century China," *Journal of Economic History,* XXVII. No. 4 (December), pp. 478–492 (1967).

Perkins, Dwight H., *Agricultural Development in China, 1368–1968.* Chicago: Aldine (1969).

Price, Derek J. de Solla, "Precision Instruments to 1500." In *A History of Technology,* Vol. III: *From the Renaissance to the Industrial Revolution, 1500–1750,* edited by Charles Singer *et al.*, pp. 582–619. London: Oxford University Press (1957).

Price, Derek J. de Solla, *Science Since Babylon.* New Haven: Yale University Press, enlarged edition (1975).

Rae, John, "The Invention of Invention" in *Technology in Western Civilization,* edited by Melvin Kranzberg and Carroll W. Pursell, Jr., Vol. I, pp. 325–336. N.Y.: Oxford University Press (1967a).

Rae, John, "Energy Conversion," in *Technology In Western Civilization,* edited by Melvin Kranzberg and Carroll W. Pursell, Jr., Vol. I, pp. 336–349. N.Y.: Oxford University Press (1967b).

Randall, Adrian J., "The Philosophy of Luddism: The Case of the West of England Workers, ca 1790–1809." *Technology and Culture,* Vol. 27, No. 1 (Jan 1986, pp. 1–18) (1986).

Reti, Ladislao, "The Double-Acting Principle in East and West," *Technology and Culture,* Vol. 11, No. 2 (April) pp. 178–200 (1970).

Reynolds, Terry S., "Scientific Influences on Technology: The Case of the Overshot Waterwheel, 1752–54," *Technology and Culture,* Vol. 20, No. 2 (April), pp.

270–95 (1979).
Reynolds, Terry S., *Stronger than a Hundred men: A History of the Vertical Water Wheel.* Baltimore; Johns Hopkins University Press (1983).
Robinson, Eric, "James Watt and the Law of Patents, *Technology and Culture,* Vol. 13, No. 2 (April), pp. 115–139 (1972).
Ronan, Colin A. and Joseph Needham, *The Shorter Science and Civilization in China,* Vol. 3. Cambridge: Cambridge University Press (1986).
Rosenberg, N., ed., *The American System of Manufactures: The Report of the Committee on the Machinery of the United States,* 1855. Edinburgh: Edinburgh University Press (1969).
Rosenberg, N., *Perspectives on Technology.* Cambridge: Cambridge University Press (1976).
Rosenberg, N., *Inside the Black Box: Technology and Economics.* Cambridge: Cambridge University Press (1982).
Rosenberg, N. and L. E. Birdzell, *How the West Grew Rich: The Economic Transformation of the Industrial World.* New York: Basic Books (1986).
Rosenberg, N. and Vincenti, Walter G., *The Britannia Bridge: The Generation and Diffusion of Technological Knowledge.* Cambridge, Mass.: M.I.T. Press (1978).
Rostoker, William, Bronson, Bennett, and Dvorak, James, "The Cast-Iron Bells of China," *Technology and Culture* Vol 25, No. 4 (Oct), pp. 750–67.
Rostow, W. W., *How it All Began.* New York: McGraw Hill (1975).
Sandberg, L. G., *Lancashire in Decline.* Columbus: Ohio University Press (1974).
Scherer, F. M., *Industrial Market Structure and Economic Performance.* Chicago: Rand McNally (1980).
Scherer, F. M., *Innovation and Growth.* Cambridge, Mass. M.I.T. Press (1984).
Schiff, Eric, *Industrialization without National Patent.* Princeton: Princeton University Press (1971).
Schubert, H. R., "Extraction and Production of Metals: Iron and Steel." In *A History of Technology,* Vol. IV: *The Industrial Revolution,* 1750–1850, edited by Charles Singer *et al.*, pp. 99–117. London: Oxford University Press (1958).
Schumpeter, Joseph A., *The Theory of Capitalist Development.* Cambridge, Mass.: Harvard University Press (1934).
Schumpeter, Joseph A., *Capitalism, Socialism and Democracy.* Third edition, New York: Harper and Row (1950).
Scott, Joan W., *The Glassworkers of Carmaux.* Cambridge: Harvard University Press (1974).
Scoville, Warren C., *Capitalism and French Glassmaking.* Berkeley: University of California Press (1950).
Sewell, William H., *Work and Revolution in France.* Cambridge: Cambridge University Press (1980).
Sharlin, Harold I. "Applications of Electricity" in *Technology in Western Civilization,* edited by Melvin Kranzberg and Carroll W. Pursell, Jr., Vol. I, pp. 563–578. N.Y.: Oxford University Press (1967a).
Sharlin, Harold I. "Electrical Generation and Transmission," in *Technology in Western Civilization,* edited by Melvin Kranzberg and Carroll W. Pursell, Jr., Vol. I, pp. 578–591, N.Y.: Oxford University Press (1967b).
Shorter, Edward and Tilly, Charles, *Strikes in France,* 1830–1968. Cambridge: Cambridge University Press (1974).
Simon, Julian, *The Economics of Population Growth.* Princeton: Princeton University Press (1977).
Simon, J., "The Effects of Population on Nutrition and Economic Well-Being." In *Hunger and History: The Impact of Changing Food Production and Consump-*

tion Patterns on Society, edited by R. I. Rotberg and T. K. Rabb, pp. 215–240. Cambridge: Cambridge University Press (1983).

Singer, Charles, "East and West in Retrospect. In *A History of Technology,* Vol. II: *The Mediterranean Civilizations and the Middle Ages, 700* B.C. *to* A.D. *1500,* edited by Charles Singer *et al.,* pp. 753–776. London. Oxford University Press.

Smith, Cyril Stanley, "Metallurgy in the Seventeenth and Eighteenth Century" in *Technology in Western Civilization,* edited by Melvin Kranzberg and Carroll W. Pursell, Jr., Vol. I, pp. 142–167; N.Y.: Oxford University Press (1967a).

Smith, Cyril Stanley, "Mining and Metallurgical Production," in *Technology in Western Civilization,* edited by Melvin Kranzberg and Carroll W. Pursell, Jr., Vol. I, pp. 349–366. N.Y.: Oxford University Press (1967b).

Smith, Cyril Stanley, "Metallurgy: Science and Practice Before 1900," *Technology in Western Civilization,* edited by Melvin Kranzberg and Carroll W. Pursell, Jr., Vol. I, pp. 592–601. N.Y.: Oxford University Press (1967c).

Smith, Cyril Stanley and Forbes, R. J., "Metallurgy and Assaying." In *A History of Technology,* Vol. III: *From the Renaissance to the Industrial Revolution, 1500–1750,* edited by Charles Singer *et al.,* pp. 27–71. London: Oxford University Press (1957).

Smith, Adam, *The Wealth of Nations,* Cannan ed. Chicago: University of Chicago Press (1976).

Smith, Norman A. F., "The Origins of the Water Turbine and the Invention of its Name," *History of Technology,* Vol. II, pp. 215–259 (1977).

Smith, Merritt Roe, *Harpers Ferry Armory and the New Technology.* Ithaca: Cornell University Press (1977).

Stern, Bernhard J., "Resistances to the Adoption of Technological Innovations," in *Technological Trends and National Policy.* Washington, D. C., United States Government Printing Office (1937).

Stevenson, John. *Popular Disturbances in England,* 1700–1870. New York: Longman (1979).

Szostak, Richard, "The Role of Transportation in the English Industrial Revolution," unpublished ms., University of Alberta (1986).

Taylor, E. G. R., *The Haven-Finding Art: A History of Navigation from Odysseus to Captain Cook.* New York: Aberlard-Schuman (1957).

Taylor, F. Sherwood and Charles Singer, "Pre-Scientific Industrial Chemistry." In *A History of Technology,* Vol. II: *The Mediterranean Civilizations and the Middle Ages,* 700 *B.C. to A.D.* 1500, edited by Charles Singer *et al.,* pp. 347–373. London: Oxford University Press (1956).

Thirtle, Colin G. and Ruttan, Vernon W., *The Role of Demand and Supply in the Generation and Diffusion of Technical Change.* Chur, Switzerland: Harwood Academic Publishers (1987).

Tsuen-Hsuin, Tsien, "Why Paper and Printing were Invented first in China and used later in Europe", In *Explorations in the History of Science and Technology in China,* edited by Li Guohao, Zhang Mengwen, and Cao Tianqin, pp. 459–470. Shanghai: Chinese Classics Publishing House (1982).

Tsuen-Hsuin, Tsien, *Paper and Printing,* in *Science and Civilization in China,* edited by Joseph Needham, Vol. 5, part I. Cambridge: Cambridge University Press (1985).

Von Tunzelmann, G. N., *Steam Power and British Industrialization to* 1860. Oxford: Oxford University Press (1978).

Von Tunzelmann, G. N., "Technical Progress during the Industrial Revolution," in *The Economic History of Britain since 1700,* edited by Roderick Floud and Donald N. McCloskey, Vol. I, pp. 143–163, Cambridge: Cambridge University

Press (1981).
Tweedale, Geoffrey, "Metallurgy and Technological Change: A Case Study of Sheffield Specialty Steel," *Technology and Culture,* Vol. 27, No. 2 (April), pp. 189–226 (1986).
Tylecote, R. F., *A History of Metallurgy.* London: The Metals Society (1976).
Unger, Richard, *Dutch Shipbuilding Before 1800.* Assen: Van Gorcum (1978).
Unger, Richard, *The Ship in the Medieval Economy, 600–1600.* London: Croom Helm (1980).
Unger, Richard, "Warships and Cargo Ships in Medieval Europe," *Technology and Culture,* Vol. 22, No. 2 (April 1981), pp. 233–52.
Usher, Abbott P., *A History of Mechanical Inventions.* Cambridge, Mass.: Harvard University Press (1954).
Usher, Abbott P., "Technical Change and Capital Formation and Economic Growth, National Bureau of Economic Research. Reprinted in *The Economics of Technological Change,* edited by Nathan Rosenberg. Harmondsworth: Penguin Books (1955).
Usher, Abbott P., "Machines and Mechanisms." In *A History of Technology,* Vol. III: *From the Renaissance to the Industrial Revolution, 1500–1750,* edited by Charles Singer *et al.,* pp. 324–346. London: Oxford University Press (1957).
Wadsworth, Alfred P. and Julia De Lacy Mann, *The Cotton Trade and Industrial Lancashire.* Manchester: Manchester University Press (1931).
Wailes, Bernard, "Plow and Population in Temperate Europe," in *Population Growth: Anthropological Implications,*" edited by Brian Spooner, pp. 154–79. Cambridge, Mass.: M.I.T. Press (1972).
Wailes, Rex, "Windmills." In *A History of Technology,* Vol. III: *From the Renaissance to the Industrial Revolution, 1500–1750,* edited by Charles Singer *et al.,* Vol. III, pp. 89–109. London: Oxford University Press (1957).
Watson, Andrew M., *Agricultural Innovation in the Early Islamic World.* Cambridge: Cambridge University Press (1983).
Wells, F. A., "Hosiery and Lace," In *A History of Technology,* Vol. V: *The Late Nineteenth Century,* edited by Charles Singer *et al.,* pp. 595–604. London: Oxford University Press (1958).
White, K. D., *Greek and Roman Technology.* Ithaca: Cornell University Press (1984).
White, Lynn, *Medieval Technology and Social Change.* Oxford: Oxford University Press (1962).
White, Lynn, *Dynamo and Virgin Reconsidered.* Cambridge, Mass.: MIT Press (1968).
White, Lynn, "The Expansion of Technology, 500–1500." In *The Fontana Economic History of Europe,* Vol. I: *The Middle Ages,* edited by C. M. Cipolla, Vol. I., 143–174. London: Collins (1972).
White, Lynn, *Medieval Religion and Technology.* Berkeley: University of California Press (1978).
Wiet, G., "The Moslem World, Seventh to Thirteenth Century." In *A History of Technology and Invention,* Vol. I: *The Origins of Technological Civilization,* edited by Maurice Daumas, pp. 336–372. New York: Crown.
Wikander, Orjan, "Archaeological Evidence for Early Water-Mills: an Interim Report," *History of Technology,* Vol. 10, pp. 1151–179 (1985).
Wilkinson, Norman B., "Brandywine Borrowings From European Technology," *Technology and Culture,* Vol. IV, No. 1, pp. 1–13 (Winter, 1963).
Williams, Martha W., "Infant Nutrition and Economic Growth in Western Europe from the Middle Ages to the Modern Period," unpublished Ph.D. dissertation,

Northwestern Univeristy (1988).

Williams, Trevor I., "Heavy Chemicals," In *A History of Technology,* Vol. V. V: *The Late Nineteenth Century,* edited by Charles Singer *et al.*, pp. 235–256. London: Oxford University Press (1958).

Woodbury, Robert S., *Studies in the History of Machine Tools.* Cambridge, Mass.: M.I.T. Press (1972).

Woronoff, Denis, *L'industrie sidérurgique en France pendant la Révolution et l'Empire.* Paris: Ecole de hautes Etudes en Sciences Sociales (1984).

Wrigley, E. A., *People, Cities and Wealth.* Oxford: Basil Blackwell (1987).

Wrigley, Julia, "Technical Education and Industry in the Nineteenth Century," in *The Decline of the British Economy,* edited by Bernard Elbaum and William Lazonick. Oxford: Clarendon Press, pp. 162–188 (1986).

Wyatt, Geoffrey, *The Economics of Invention: A Study of the Determinants of Inventive Activity.* New York: St. Martin's (1986).

Young, Allyn, "Increasing Returns and Economic Progress," *Economic Journal,* Vol. 38, No. 4 (December) (1928).

INDEX

FUNDAMENTALS OF PURE AND APPLIED ECONOMICS

SECTIONS AND EDITORS

BALANCE OF PAYMENTS AND INTERNATIONAL FINANCE
W. Branson, Princeton University

DISTRIBUTION
A. Atkinson, London School of Economics

ECONOMIC DEMOGRAPHY
T.P. Schultz, Yale University

ECONOMIC DEVELOPMENT STUDIES
S. Chakravarty, Delhi School of Economics

ECONOMIC FLUCTUATIONS: FORECASTING, STABILIZATION, INFLATION, SHORT TERM MODELS, UNEMPLOYMENT
A. Ando, University of Pennsylvania

ECONOMIC HISTORY
P. David, Stanford University, and M. Lévy-Leboyer, Université Paris X

ECONOMIC SYSTEMS
J.M. Montias, Yale University, and J. Kornai, Institute of Economics, Hungarian Academy of Sciences

ECONOMICS OF HEALTH, EDUCATION, POVERTY AND CRIME
V. Fuchs, Stanford University

ECONOMICS OF THE HOUSEHOLD AND INDIVIDUAL BEHAVIOR
J. Muellbauer, University of Oxford

ECONOMICS OF TECHNOLOGICAL CHANGE
F. M. Scherer, Swarthmore College

ECONOMICS OF UNCERTAINTY AND INFORMATION
S. Grossman, Princeton University, and J. Stiglitz, Princeton University

EVOLUTION OF ECONOMIC STRUCTURES, LONG-TERM MODELS, PLANNING POLICY, INTERNATIONAL ECONOMIC STRUCTURES
W. Michalski, O.E.C.D., Paris

EXPERIMENTAL ECONOMICS
C. Plott, California Institute of Technology

GAME THEORY
R. Aumann, The Hebrew University of Jerusalem

GENERAL EQUILIBRIUM THEORY AND OPTIMUM THEORY
W. Hildenbrand, University of Bonn, and A. Mas-Colell, Harvard University

GOVERNMENT OWNERSHIP AND REGULATION OF ECONOMIC ACTIVITY
E. Bailey, Carnegie-Mellon University

INTERNATIONAL ECONOMIC ISSUES
T. Fujii, University of Nagoya

INTERNATIONAL TRADE
M. Kemp, University of New South Wales

LABOR ECONOMICS
F. Welch, University of California, Los Angeles

LAW AND ECONOMICS
S. Shavell, Harvard Law School

MACROECONOMIC THEORY
J. Grandmont, CEPREMAP

MARXIAN ECONOMICS
J. Roemer, University of California, Davis

MONETARY THEORY
N. Wallace, University of Minnesota

NATURAL RESOURCES AND ENVIRONMENTAL ECONOMICS
C. Henry, Ecole Polytechnique, Paris

ORGANIZATION THEORY AND ALLOCATION PROCESSES
A. Postlewaite, University of Pennsylvania, and D. Schmeidler, Tel Aviv University

POLITICAL SCIENCE AND ECONOMICS
J. Ferejohn, Stanford University

PROGRAMMING METHODS IN ECONOMICS
M. Balinski, Ecole Polytechnique, Paris

PUBLIC EXPENDITURES
P. Dasgupta, University of Oxford

REGIONAL AND URBAN ECONOMICS
R. Arnott, Queen's University at Kingston

SOCIAL CHOICE THEORY
A. Sen, University of Oxford

TAXES
R. Guesnerie, Ecole des Hautes Etudes en Sciences Sociales

THEORY OF ECONOMIC GROWTH
J. Scheinkman, University of Chicago

THEORY OF THE FIRM AND INDUSTRIAL ORGANIZATION
A. Jacquemin, Université Catholique de Louvain

PUBLISHED TITLES

Volume 1 (International Trade Section)
GAME THEORY IN INTERNATIONAL ECONOMICS
by John McMillan

Volume 2 (Marxian Economics Section)
MONEY, ACCUMULATION AND CRISIS
by Duncan K. Foley

Volume 3 (Theory of the Firm and Industrial Organization Section)
DYNAMIC MODELS OF OLIGOPOLY
by Drew Fudenberg and Jean Tirole

Volume 4 (Marxian Economics Section)
VALUE, EXPLOITATION AND CLASS
by John E. Roemer

Volume 5 (Regional and Urban Economics Section)
LOCATION THEORY
by Jean Jaskold Gabszewicz and Jacques-François Thisse, Masahisa Fujita, and Urs Schweizer

Volume 6 (Political Science and Economics Section)
MODELS OF IMPERFECT INFORMATION IN POLITICS
by Randall L. Calvert

Volume 7 (Marxian Economics Section)
CAPITALIST IMPERIALISM, CRISIS AND THE STATE
by John Willoughby

Volume 8 (Marxian Economics Section)
MARXISM AND "REALLY EXISTING SOCIALISM"
by Alec Nove

Volume 9 (Economic Systems Section)
THE NONPROFIT ENTERPRISE IN MARKET ECONOMIES
by Estelle James and Susan Rose-Ackerman

Volume 10 (Regional and Urban Economics Section)
URBAN PUBLIC FINANCE
by David E. Wildasin

Volume 11 (Regional and Urban Economics Section)
URBAN DYNAMICS AND URBAN EXTERNALITIES
by Takahiro Miyao and Yoshitsugu Kanemoto

Volume 12 (Marxian Economics Section)
DEVELOPMENT AND MODES OF PRODUCTION IN MARXIAN ECONOMICS: A CRITICAL EVALUATION
by Alan Richards

Volume 13 (Economics of Technological Change Section)
TECHNOLOGICAL CHANGE AND PRODUCTIVITY GROWTH
by Albert N. Link

Volume 14 (Economic Systems Section)
ECONOMICS OF COOPERATION AND THE LABOR-MANAGED ECONOMY
By John P. Bonin and Louis Putterman

Volume 15 (International Trade Section)
UNCERTAINTY AND THE THEORY OF INTERNATIONAL TRADE
by Earl L. Grinols

Volume 16 (Theory of the Firm and Industrial Organization Section)
THE CORPORATION: GROWTH, DIVERSIFICATION AND MERGERS
by Dennis C. Mueller

Volume 17 (Economics of Technological Change Section)
MARKET STRUCTURE AND TECHNOLOGICAL CHANGE
by William L. Baldwin and John T. Scott

Volume 18 (Social Choice Theory Section)
INTERPROFILE CONDITIONS AND IMPOSSIBILITY
by Peter C. Fishburn

Volume 19 (Macroeconomic Theory Section)
WAGE AND EMPLOYMENT PATTERNS IN LABOR CONTRACTS: MICROFOUNDATIONS AND MACROECONOMIC IMPLICATIONS
by Russell W. Cooper

Volume 20 (Government Ownership and Regulation of Economic Activity Section)
DESIGNING REGULATORY POLICY WITH LIMITED INFORMATION
by David Besanko and David E. M. Sappington

Volume 21 (Economics of Technological Change Section)
THE ROLE OF DEMAND AND SUPPLY IN THE GENERATION AND DIFFUSION OF TECHNICAL CHANGE
by Colin G. Thirtle and Vernon W. Ruttan

Volume 22 (Regional and Urban Economics Section)
SYSTEMS OF CITIES AND FACILITY LOCATION
by Pierre Hansen, Martine Labbé, Dominique Peeters and Jacques-François Thisse, and J. Vernon Henderson

Volume 23 (International Trade Section)
DISEQUILIBRIUM TRADE THEORIES
by Motoshige Itoh and Takashi Negishi

Volume 24 (Balance of Payments and International Finance Section)
THE EMPIRICAL EVIDENCE ON THE EFFICIENCY OF FORWARD AND FUTURES FOREIGN EXCHANGE MARKETS
by Robert J. Hodrick

Volume 25 (Economic Systems Section)
THE COMPARATIVE ECONOMICS OF RESEARCH DEVELOPMENT AND INNOVATION IN EAST AND WEST: A SURVEY
by Philip Hanson and Keith Pavitt

Volume 26 (Regional and Urban Economics Section)
MODELING IN URBAN AND REGIONAL ECONOMICS
By Alex Anas

Volume 27 (Economic Systems Section)
FOREIGN TRADE IN THE CENTRALLY PLANNED ECONOMY
by Thomas A. Wolf

Volume 28 (Theory of the Firm and Industrial Organization Section)
MARKET STRUCTURE AND PERFORMANCE – THE EMPIRICAL RESEARCH
by John S. Cubbin

Volume 29 (Economic Development Studies Section)
STABILIZATION AND GROWTH IN DEVELOPING COUNTRIES: A STRUCTURALIST APPROACH
by Lance Taylor

Volume 30 (Economics of Technological Change Section)
THE ECONOMICS OF THE PATENT SYSTEM
by Erich Kaufer

Volume 31 (Regional and Urban Economics Section)
THE ECONOMICS OF HOUSING MARKETS
by Richard F. Muth and Allen C. Goodman

Volume 32 (International Trade Section)
THE POLITICAL ECONOMY OF PROTECTION
by Arye L. Hillman

Volume 33 (Natural Resources and Environmental Economics Section)
NON-RENEWABLE RESOURCES EXTRACTION PROGRAMS AND MARKETS
by John M. Hartwick

Volume 34 (Government Ownership and Regulation of Economic Activity Section)
A PRIMER ON ENVIRONMENTAL POLICY DESIGN
by Robert W. Hahn

Volume 35 (Economics of Technological Change Section)
TWENTY-FIVE CENTURIES OF TECHNOLOGICAL CHANGE
by Joel Mokyr

Volume 36 (Government Ownership and Regulation of Economic Activity Section)
PUBLIC ENTERPRISE IN MONOPOLISTIC AND OLIGOPOLISTIC INDUSTRIES
by Ingo Vogelsang

Further titles in preparation
ISSN: 0191-1708

For Product Safety Concerns and Information please contact our EU representative GPSR@taylorandfrancis.com
Taylor & Francis Verlag GmbH, Kaufingerstraße 24, 80331 München, Germany

www.ingramcontent.com/pod-product-compliance
Lightning Source LLC
LaVergne TN
LVHW021111300926
849160LV00009B/686

* 9 7 8 3 7 1 8 6 4 9 3 6 5 *